1987

Peptide Hormones, Biomembranes, and Cell Growth

Peptide Hormones, Biomembranes, and Cell Growth

Edited by
Liana Bolis

Institute of General Physiology
Messina, Italy

Roberto Verna
and
Luigi Frati

Institute of General Pathology
Rome, Italy

Plenum Press • New York and London

Library of Congress Cataloging in Publication Data

International Meeting on Peptide Hormones, Biomembranes, and Cell Growth (1983: Rome, Italy)
 Peptide hormones, biomembranes, and cell growth.

"Proceedings of an International Meeting on Peptide Hormones, Biomembranes, and Cell Growth, held October 12–14, 1983, in Rome, Italy"—T.p. verso.
 Includes bibliographies and index.
 1. Peptide hormones—Congresses. 2. Cell membranes—Congresses. 3. Cells—Growth—Congresses. I. Bolis, Liana. II. Verna, Roberto. III. Frati, Luigi. IV. Title. [DNLM: 1. Peptides—metabolism—congresses. 2. Hormones—metabolism—congresses. 3. Cell Membrane—metabolism—congresses. 4. Membrane Fluidity—congresses. 5. Receptors, Endogenous Substances—metabolism—congresses. WK 102 I606p 1983]
QP572.P4I58   1983   599′.01927   84-17846
ISBN 0-306-41816-9

Proceedings of an International Meeting on Peptide Hormones, Biomembranes, and Cell Growth, held October 12–14, 1983, in Rome, Italy

©1984 Plenum Press, New York
A Division of Plenum Publishing Corporation
233 Spring Street, New York, N.Y. 10013

Printed in the United States of America

PREFACE

The field of study of receptors today is one of the most innovative in biology and pathology.

This book provides a multidisciplinary approach to the problem: the biochemical characterization of binding, the mode of action of receptors, their relationship to ion transport as well as the relevance of membrane fluidity in receptor activity are discussed.

It is hoped that this volume will stimulate further collaboration among scientists in both basic and applied disciplines.

The thanks of all the participants are particularly due to Roberto Verna, who, with his efficiency and enthusiasm, has organized such an outstanding scientific meeting.

Felice G. Caramia M.D.
Professor of General Pathology
University of Rome, Italy

v

CONTENTS

CONTENTS

INTRODUCTION

C. Liana Bolis

Istituto di Fisiologia Generale
Universita di Messina
98100 Messina
Italy

The importance of biomembranes as the recognition site for
many hormones, neurotransmitters, toxins and drugs is now well
documented. In addition, cell cycle steady states in actively
growing cells seem to play an important role in the control of
receptor density (Holley et al., 1977). For this reason it seems
particularly important today to consider cellular activity as an
essential part of the cooperativity of membranes in signal
recognition and transduction (Bolis and Luly, 1978). For
example, cell receptor numbers can be observed to change during
cell cycle, cellular development and differentiation.

During the 1950's various experimental approaches were
developed to study hormonal binding to target cells; later,
radiolabelled hormones with high specific activities made
available information on the binding characteristics and the
specificity of hormones for specific receptor sites (Cuatrecasas
and Hollenberg 1976). These studies have been conducted with
intact cell preparations, purified or crude hormone preparations,
and solubilized receptors (Kono, 1969; Roth, 1973). However, the
experimental approaches all lack the objectivity of a study which
takes into account simultaneously overall cellular activity as
occurs in vivo. This is true even if we know that the timeless
sequence of the binding induces a relevant response through
physiological changes in the cell, involving plasma membrane
linked phenomena, like adenylate cyclase activation and/or
permeability changes (Haynes et al. 1960; Jarett and Smith, 1974).

1

Considerable progress in the understanding of the membrane structural organization in time and space has been achieved (Singer, 1976; Singer and Nicholson, 1972). The evidence that certain membrane proteins are free to diffuse in the plane of the plasma membrane allows one to consider the possibility that hormone receptors may also be mobile entities that interact with other molecules in the membrane's plane. This concept of mobile receptors developed by Jacobs and Cuatrecasas, 1976; De Haen, 1976; Boeynaems and Dumont, 1977, is based on observations that receptor-specific agonists (AcTH, prostaglandins, glucagon, catecholamines, etc.) independently can stimulate adenylate cyclase in the adipocyte; this offers a new approach to the multireceptors concept.

Receptors at all surface levels may be affected by several conditions secondary to intracellular events (cell cycle, cell differentiation and rate of synthesis and turnover) or by a variety of external stimuli determined by hormones and other agents (homospecific or heterospecific) or chemical toxins or virus (Hollenberg, 1979).

In this meeting, very interesting aspects of the role of membrane fluidity which relates to the mobility of molecules in the plane of the membranes, were discussed, as well as the kinetics regulating receptor binding events, and other molecular events underlying the hormonal effects.

Even more important is the pathology of hormone/receptor interaction, mainly due to both hormone and receptor concentration changes, alteration in affinities, as well as anti-receptor antibodies pathology (Roth and Taylor, 1982).

REFERENCES

Boeynaems, J. M. and Dumont, J. E., 1977, The Two-Step Model of Ligand-receptor Interaction. Mol. Cell Endocrinol., 7:33

Bolis, L., and Luly, P., 1978, Membrane Receptors, in: "First Symposium on Organ Specific Autoimmunity", P. Miescher, ed., Schwabe, Basle, p. 127

Cuatrecasas, P. and Hollenberg, M. D., 1976, Membrane Receptors and Hormone Action. Advan. Protein Chem., 30:251

Dehaen, C., 1976, The Non-stoichiometric Floating Receptor Model for Hormone-sensitive Adenylate Cyclase. J. Theor. Biol., 58:383

Haynes, R. C. et al., 1960, The Role of Cyclic Adenylic Acid in Hormone Action. Recent Progr. Hormone Res., 16:121

Hollenberg, M. D., 1979, Hormone Receptor Interactions at the
    Cell Membrane. <u>Pharmacological Reviews</u>, Vol. 30, No. 4

Holley, R. W. et al., 1977, Density-dependent Regulation of Growth
    of BSC-1 Cell Culture: Control of Growth by Serum Factors.
    <u>Proc. Nat. Acad. Sci.</u>, 74:5046

Jacobs, S. and Cuatrecasas, P., 1976, The Mobile Receptor
    Hypothesis and "Cooperativity" of Hormone Binding Application
    to Insulin. <u>Biochem. Biophys. Acta</u>, 433:482

Jarett, L. and Smith, R. M., 1974, The Stimulation of Adipocyte
    Plasma Membrane Magnesium Ion-Stimulated Adenosine
    Triphosphatase by Insulin and Concanavalin A., <u>J. Biol.
    Chem.</u>, 249:5195

Kono, T., 1969, Destruction and Restoration in the Insulin
    Effector System in Isolated Fat Cells. <u>J. Biol. Chem.</u>,
    244:5777

Roth, J., 1973, Peptide Hormone Binding to Receptors: A Review of
    Direct Studies <u>In Vitro</u>. <u>Metabolism</u>, 22:1059

Roth, J. and Taylor, S. I., 1982, Receptors for Peptide Hormones:
    Alterations in Diseases of Humans. <u>Ann. Rev. Physiol.</u>, 44:639

Singer, S. J., 1976, The Fluid Mosaic Model of Membrane Structure
    Some Applications to Ligand-receptor and Cell-cell
    Interactions, <u>in</u>: "Surface Membrane Receptors", R. A.
    Bradshaw, W. A. Frazier, R. C. Merrell, D. I. Gottlieb and
    Hogue-Angeletti, R. A., eds., Plenum Press, New York

Singer, S. J. and Nicholson, G. L., 1972, The Fluid Mosaic Model
    of the Structure of Cell Membranes. <u>Science</u>, 175:720

WHAT BINDING EXPERIMENTS CAN AND CANNOT TELL US ABOUT THE

INTERACTION BETWEEN HORMONES AND MEMBRANES

Allen P. Minton

Laboratory of Biochemical Pharmacology, National
Institute of Arthritis, Diabetes, Digestive and Kidney
Diseases, N.I.H., Bethesda, MD 20205 U.S.A.

INTRODUCTION

Until less than twenty years ago, information about the interaction between a hormone and its target tissue had to be indirectly
inferred from the dependence of the magnitude of elicited response
upon hormone dose or concentration. Following development of methods for the preparation of radiolabeled hormones of high specific
activity (reviewed in Cuatrecasas and Hollenberg, 1976) it became
possible to measure directly the binding of as little as a few
picomoles of hormone to a target preparation. By correlating the
extent of binding with the dose and with the extent of elicited
response it became possible to identify, and in some cases, isolate,
particular sets of binding sites termed receptors, the occupation
of which by hormone is associated with (and assumed to be) the
initiating event in the elicitation of response by hormone.

The ability of receptors to specifically bind a particular
ligand has been exploited, on one hand, to facilitate the chemical
isolation and characterization of the receptor molecule, and on
the other hand, to explore the molecular mechanism by which binding
of hormone to receptor leads to the observed cellular response.
The purpose of this communication is to briefly review factors
which can affect the binding of hormones to their receptors and to
emphasize that binding data alone do not permit one to identify
which of the possible contributory factors are actually operative
in the system under investigation.

For the purpose of this work we shall use the term "binding"
to refer exclusively to a specific, saturable interaction between
hormone and one or more sets of binding sites. Experimental data

are assumed to have been corrected properly for the presence of weak, unsaturable adsorption of hormone to target tissue (commonly referred to as "nonspecific binding").

CLASSIFICATION OF BINDING ISOTHERMS

The functional dependence of binding upon ligand concentration at a single temperature is referred to as a binding isotherm. We shall classify isotherms on the basis of the appearance of the plotted data (Minton, 1982a).

The reference isotherm is that relation generated by the Langmuir equation

$$H_b/H_b^{max} = K [H] / (1 + K [H])$$ (1)

where H_b is the amount or concentration of bound hormone, H_b^{max} is the amount of bound hormone at saturation, [H] is the concentration of free (unbound) hormone, and K is an equilibrium constant for association of univalent hormone with a single class of independent, equivalent binding sites. The titration plot of the reference isotherm (Figure 1a, solid curve) is sigmoid, symmetrical about the half-saturation point, with two log units in free hormone concentration separating the 9 per cent and 91 per cent levels of site saturation. The Scatchard plot of the reference isotherm (Figure 1b, solid curve) is a straight line.

Two other general types of isotherm frequently observed are schematically depicted in Figures 1a and 1b. Curves resembling those drawn with short dashes in the figures will be referred to as apparent cooperative isotherms. Less than two log units in free hormone concentration separate the 9 per cent and 91 per cent levels of site saturation in a titration plot, and the Scatchard plot is convex upward. Curves resembling those drawn with alternate dots and dashes in the figures will be referred to as apparent multiple site class isotherms. More than two log units in free hormone concentration separate the 9 per cent and 91 per cent levels of site saturation in a titration plot, and the Scatchard plot is concave upward.

FACTORS WHICH MAY INFLUENCE THE BINDING OF HORMONE TO MEMBRANE RECEPTORS

While the interpretation of a reference-type isotherm in terms of a single homogenous (or quasi-homogeneous) class of independent binding sites is straightforward, both the apparent multiple site and apparent cooperative types of isotherms may result from a variety of underlying causes. These are listed for reference in Table I and briefly discussed below.

1. If binding sites for hormone are partitioned into two or more discrete, independent classes of significantly different affinity for hormone, an apparent multiple-site class isotherm will result (Scatchard, 1949).

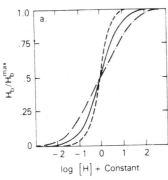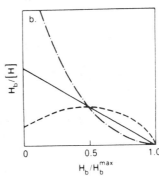

Fig. 1.   Titration plot (a) and Scatchard plot (b) of three phenomenological types of binding isotherms. Reference isotherm, ———— ; apparent cooperative isotherm —————— ; apparent multiple site class isotherm, –·–·–·–·. (Adapted from Minton, 1982a).

Table 1. Possible interpretations of isotherm shape.

Reference	Single homogeneous class of independent binding sites "Quasi-homogeneous" distribution of site affinity (Minton, 1979)
Apparent multiple site class	Multiple discrete classes of independent binding sites (Scatchard, 1949) Continuous distribution of site affinity (Karush and Sonnenberg, 1949) Negatively cooperative interaction between binding sites or hormone molecules (Scatchard, 1949) Nonequilibrium effects (Minton, 1982b) Nonstoichiometric receptor-effector interaction (Jacobs and Cuatrecasas, 1976)
Apparent cooperative	Positively cooperative interaction between binding sites or hormone molecules (Scatchard, 1949) Nonequilibrium effects (Minton, 1982b)

2. If binding sites for hormone consist of a group of closely related but nonidentical molecules (such as a glycoprotein of variable carbohydrate composition), or if the local environment of a receptor molecule varies from point to point over the membrane surface, a distribution of affinity for hormone may result, together with a correspondingly broadened (multiple site type) isotherm (Karush and Sonnenberg, 1949).

3. The affinity of a particular receptor molecule for hormone may depend upon the occupancy state of neighboring receptor molecules. If the occupancy of a receptor molecule by hormone decreases the affinity of a neighboring receptor molecule for hormone, a negatively cooperative interaction is said to exist between receptor molecules, and the binding isotherm will be broadened. If the occupancy of a receptor molecule by hormone increases the affinity of a neighboring receptor molecule for hormone, a positively cooperative interaction is said to exist between receptor molecules, and the binding isotherm will be steepened (Scatchard, 1949).

4. Cooperative interactions may exist between hormone molecules as well as between receptor molecules. One type of negatively cooperative interaction could result from the clustering of receptors into patches of high surface density. If the hormone molecule were large enough, simple steric repulsion between hormone molecules might prevent complete saturation of all binding sites in a patch, and the resulting binding isotherm would be greatly broadened.

5. If binding of hormone alters the rate of a reaction taking place continuously during the binding assay, a steady state far from equilibrium may be achieved in which the hormone binding isotherm may be either steepened or broadened (Minton, 1982b).

6. If binding of hormone to receptor alters the interaction between receptor and another macromolecular component of the target membrane (called effector), and if effector is present in a stoichiometric ratio to receptor less than unity, then an apparent two site class isotherm may result, reflecting the difference in affinity for hormone between receptor which interacts with effector, and receptor which is in excess of effector and cannot interact with it (Jacobs and Cuatrecasas, 1976).

It should be evident at this point that one cannot discriminate between the various alternatives offered above on the basis of binding studies carried out in tissue or membrane homogenates alone. However, a combination of binding measurements carried out on "intact" membranes and upon purified or partially purified membrane components may permit the range of choice to be substantially narrowed. An example of such a combined study may be found in Maturo and Hollenberg (1978).

INTERPRETING THE COMPETITIVE BINDING ASSAY: A CAUTIONARY NOTE

The competitive binding assay provides a convenient method for comparing the binding properties of a native hormone and chemically related substances, such as hormone analogs, which bind to the same receptors. Under the proper conditions the competitive assay may also provide quantitative binding isotherms for hormones which cannot be themselves radiolabeled without substantial loss of native binding or functional properties.

A conventional competition assay is carried out as follows (Campfield, 1983). A tracer substance (denoted here by B) is selected on the basis of two criteria: it may be radiolabeled to high specific activity, and it is displaced from its specific binding sites by a sufficiently high concentration of the unlabeled competitor (denoted by A). A series of assay samples are prepared containing fixed amounts of target tissue or membrane and tracer and varying amounts of the test substance A. The amount of bound B (B_b) is measured as a function of the concentration of unlabeled competitor ($[A]$). It is usually assumed that y_A, the fractional saturation of binding sites by A, is given by

$$y_A = A_b / A_b^{max} = 1 - B_b / B_b^o \qquad (2)$$

where B_b^o is the amount of B bound in the absence of A.

We would like to emphasize here two points which seem to have been generally neglected when applying eqn (2) to the analysis of competition data.

The first point is that the binding of A to membrane may not be adequately represented by the single species AR (where R denotes receptor). In the most general case, A may conceivably form a variety of complexes with R which may involve additional membrane components S, T, ... as well. Each of these complexes may be denoted by the general form $A_m R_n S_p T_q$ ..., where $m, n \geq 1$ and $p, q, ... \geq 0$. It is possible to show rigorously (A. Minton, unpublished calculations) that eqn (2) can only be valid if $m = n$ for each and every species of complex involving A and R. In particular, eqn (2) cannot be validly used if complexes of the form AR_2 and/or A_2R constitute a significant fraction of bound A.

The second point generally overlooked in the application of eqn (2) is that even when it is valid, the binding isotherm derived for A is only equal to that obtained via a direct binding assay when $B_b^o \ll B_b^{max}$, that is, when the fractional saturation of binding sites by B is negligible. If the fractional saturation of binding sites by B is non-negligible, then A_b (and y_A) are

functions of both [A] and [B], and may have values which are
significantly less than those obtained for the same value of [A]
in the absence of B.

REFERENCES

Campfield, L. A., 1983, Mathematical analysis of competitive pro-
     tein binding assays, in: "Principles of Competitive Protein
     Binding assays", 2nd ed., W. D. Odell and P. Franchimont,
     eds., Wiley and Sons, New York.
Cuatrecasas, P. and Hollenberg, M. D., 1976, Membrane receptors in
     hormone action, Advances in Protein Chem. 30:252.
Jacobs, S. and Cuatrecasas, P., 1976, The mobile receptor hypo-
     thesis and 'cooperativity' of hormone binding, Biochim.
     Biophys. Acta 433:482.
Karush, F. and Sonnenberg, M., 1949, The interaction of homologous
     alkyl sulfates with bovine serum albumin, J. Am. Chem. Soc.
     71:1369.
Maturo, J. M. III and Hollenberg, M. D., 1978, Insulin receptor:
     interaction with nonreceptor glycoprotein from rat liver
     membranes. Proc. Natl. Acad. Sci. U.S.A. 75:3070.
Minton, A. P., 1979, Apparent homogeneity in some heterogeneous
     systems, Biochim. Biophys. Acta 558:179.
Minton, A. P., 1982, Steady state relations between hormone binding
     and elicited response: quantitative mechanistic models, in:
     "Hormone Receptors", L. D. Kohn, ed., Wiley and Sons, New York.
Minton, A. P., 1982b, Elicitation of cellular response by bioactive
     ligands: nonequilibrium effects, Biochem. Biophys. Res.
     Commun. 107:1206.
Scatchard, G., 1949, The attraction of proteins for small molecules
     and ions, Ann. N.Y. Acad. Sci. 51:660.

THE BINDING OF MULTIVALENT EFFECTOR TO MEMBRANES AND ITS KINETIC

REGULATION

Roberto Strom

Department of Human Biopathology
University of Rome "La Sapienza"
Rome, Italy

INTRODUCTION

Transmembrane signaling is often triggered by the clustering of two or more receptors on the cell membrane (1,2). In basophils or mast cells, for example, cross-linking of membrane receptors, such as can be induced by multivalent ligands but not by monovalent ones, is mandatory both for histamine release and for cell desensitization, the response of the basophil depending therefore on a balance between stimulatory and inhibitory signals (3-9). Similar phenomena are also observed in lymphocyte stimulation: only tetravalent lectins can induce blastogenesis (10), and, similarly, antigen multivalency -- or at least, divalency -- is required for stimulation of B cells (11-17). A cluster of 10-20 receptors, an "immunon", has been hypothesized as requisite for stimulation of B lymphocytes in the absence of T cells (18,19). Tominaga, Takatsu and Hamaoka (20) have also shown that differentiation of B lymphocytes into antibody-secreting cells can be obtained in the absence of T cells (or of T cell replacing factors) by using a bivalent antibody capable of cross-linking the cell surface receptors which recognize the T cell replacing factor. Cluster formation is on the other hand apparently also a necessary step in the induction of B cell tolerance (13,20), such a condition being possibly due to formation of clusters too small to be "immunons".

Other biological events involving transmission of a signal across a membrane, such as phagocytosis and chemotaxis, are also likely to utilize receptor clustering as a component of the signaling process (21). Similar evidence shows that also the action of some peptide hormones on target cells can depend on the cross-linking of membrane receptors for these hormones (1,22). Thus, for

example, while both bivalent antibodies and monovalent Fab fragments
of autoantibodies directed against the insulin receptor are competitive antagonists of insulin binding, only the bivalent antibodies
stimulate the transport and oxidation of glucose (23,24). Monovalent
antibodies are inactive, but their bio-activity can be restored by
addition of a second antibody.

In many systems, therefore, transmembrane signal generation
cannot be a consequence of the overall occupancy of receptors by
ligands, but, rather, of receptor clustering, the ligand itself
being even not necessary if other means, such as bivalent antibodies,
are available to induce such a clustering. It appears therefore to
be of some interest to examine whether and to what extent independent binding, to cell membrane receptors, of equivalent determinants
belonging to a same multivalent effector can account for the experimentally observable dose-response curves. Further developments
will concern the possibility that the binding of these multivalent
effectors to the membrane receptors lead to closed ring formation,
thus introducing some deviation from the hypothesis of absolute
equivalence and independence of the single effector sites. In the
last section, a more realistic kinetic dimension will be added to
the previous quasi-equilibrium approach, some experimental results
in model systems indicating that a relevant role could be played,
in the actual regulation of multivalent binding, by the rate of
lateral diffusion of the receptors in the plane of the cell membrane.

RECEPTOR BINDING OF MULTIVALENT EFFECTORS HAVING EQUIVALENT AND IN-
DEPENDENT SITES

For the binding of a multivalent effector to cell membrane
(univalent) receptors, we may assume (25,26) that:
1) binding occurs by steps in which single bonds are formed or dis-
   rupted. Apart from the first one, successive bonds of an effector molecule to the receptor sites are allowed and regulated by
   lateral diffusion of the receptors and of the receptor-effector
   complexes;
2) with respect to their ability of reciprocal interaction, effector
   sites and receptors are considered to be all equivalent and mu-
   tually independent;
3) the probability of dissociation, within a finite time interval,
   is non-zero for any bond and, for a multiply bound molecule, is
   the same for all bonds.

Let v then be the total number of sites on an effector molecule;
f the effective valence of the effector with respect to the surface
binding, i.e. the maximum number of coexisting bonds that an effector molecule may establish at the cell surface with the receptors.
Let S_o be the total number of receptors per cell and $S(t)$ the number of unoccupied receptors per cell at time t, $c(t)$ the molar concentration of free effector in solution, $C_i(t)$ (with i=1,2,...,f)

the number of effector molecules per cell linked at the cell surface by i bonds. The binding process to the cell surface can be dealt with by a mass action law model (26):

$$\frac{dC_1}{dt} = k_{a_1}c(t)S(t) - k_{d_1}C_1(t) + k_{d_2}C_2(t) - k_{a_2}C_1(t)S(t)$$

.
.
.

$$\frac{dC_j}{dt} = k_{a_j}C_{j-1}(t)S(t) - k_{d_j}C_j(t) + k_{d_{j+1}}C_{j+1}(t) - k_{a_{j+1}}C_j(t)S(t)$$

.
.
.

$$\frac{dC_f}{dt} = k_{a_f}C_{f-1}(t)S(t) - k_{d_f}C_f(t), \qquad j=2,3 \ldots,f-1$$

$$S(t) = S_o - \sum_{i=1}^{f} iC_i(t)$$

in which the association and dissociation rate constants k_{a_i} and k_{d_i} are given by:

$$k_{a_i} = \begin{cases} vk_a & i = 1 \\ (f-1+1)k'_a & i = 2,3,\ldots,f \end{cases}$$

$$k_{d_i} = \begin{cases} k_d & i = 1 \\ ik'_a & i = 2,3,\ldots,f \end{cases}$$

where k_a and k_d are the on and off rate constants for the first binding of a given effector molecule to the cell receptors, while k'_a and k'_d are the rate constants characterizing the binding of successive effector sites.

At equilibrium — when c(t) is constant and equal to c — if we indicate by $\overline{C}_i$ and $\overline{S}$ the equilibrium values of the corresponding variables $C_i(t)$ and $S(t)$, we have:

$$\overline{C}_i = (v/f)cK\binom{f}{i}K'^{i-1}\overline{S}^i \qquad S \quad i=1,2,\ldots,f$$

where $K = k_a/k_d$ and $K' = k'_a/k'_d$. $\overline{S}$ is obtained by solving the equation:

$$S_o - (f+vcK)\overline{S} - (v/f)cK \sum_{i=2}^{f} i\binom{f}{i}K'^{i-1} \overline{S}^i = 0$$

Fig. 1 shows how the quantity

$$\overline{S}_m = \sum_{i=2}^{f} i\overline{C}_i ,$$

i.e. the number of surface receptors occupied at equilibrium by multiply bound effector molecules, depends on the effector valence f, on the free effector concentration $\overline{c}$ and on the equilibrium constants K and K'.

It can easily be seen:
a) that, for a relatively high value of the dissociation rate constant of multiple bonds, i.e. low K' value, the ability of the effector to form a cross-linking pattern on the cell surface (and therefore, presumably, to induce cell stimulation) is preserved, at low effector concentrations, only if the effector itself has a high number of sites;
b) conversely, that effectors with a high valence have a net gain over those with a low number of sites only if the K' value is low (high dissociation constant for multiple bonds) and if the free effector has a low concentration. It can be noticed however that the correlation between effector valence and gain in forming the surface lattice tends to a plateau as the number of sites per effector molecule is increased. If we introduce this approach in an overall model for cell stimulation (whereby the formation of the surface lattice is directly correlated to cell stimulation or differentiation, these peculiarities lead to a low responsiveness of the whole system towards variations of the concentration of high-valence effectors (26).

PREFERENTIAL FORMATION OF CLOSED RINGS AT THE CELL SURFACE

Unless the receptors are strictly univalent and completely free to move in the plane of the cell membrane, a multivalent effector can be expected to react preferably with contiguous receptors, thus going against the previous hypothesis of independence between sites of a same receptor molecule.

Fig. 2 illustrates the very simple case of a bivalent antibody combining, in solution, with a bivalent antigen: linear chains of various size can be expected to be in equilibrium with the corresponding rings, ring closure occurring by a monomolecular reaction within the same antigen-antibody complex, i.e. by a process which is quite distinct from the preceding bimolecular bond formation.

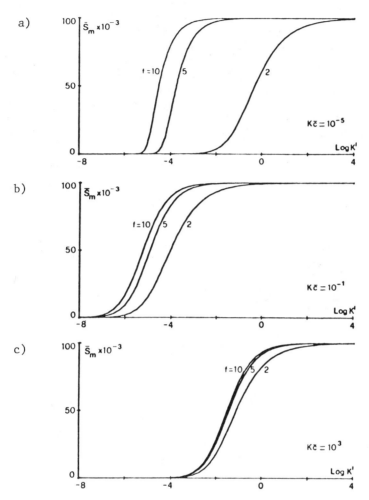

Fig. 1. Number $\bar{S}_m$ of receptor sites occupied at equilibrium by multiply bound effector molecules, as a function of Log K', for
various values of the product $K\bar{c}$, under the assumption of
independence and equivalence of sites belonging to a same
effector molecule. In all plots $S_o = 10^5$. The effective valence f is indicated on the single curves; in all cases, v/f=2
(reproduced from ref. 25).

Under the assumptions of bivalency of both antigen and antibody,
and of homogeneity of the intrinsic affinity values, a theory which
takes into account ring formation has been developed by Perelson
and DeLisi (27). Gandolfi and Strom, although using a simplified
version of this theory, have extended it (28) to the case in which
steric constraints reduce or abolish the possibility of having "monogamous" complexes (i.e. the closed ring in the top row of fig. 2)
- a fact which can be simply expressed by a low value for the monomolecular ring closure constant $K'(1)$; larger rings can instead
be formed and, to a first approximation, it can be assumed that
for these larger rings every ratio $K'(n)$ between the rate constants
for n-ring closure and for dissociation of any single bond in the
n-ring can be taken equal to $K'(n) = \overline{K}'n^{3/2}$ (29), where the value
of $\overline{K}'$ becomes independent of the size of the ring.

This same approach can be extended to ring formation on the
cell surface. A rather interesting outcome of this investigation

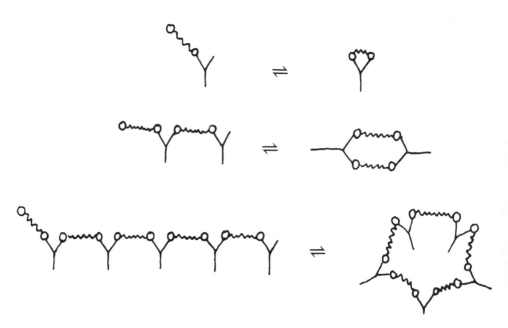

Fig. 2. Transformation of linear chains into closed rings upon interaction of a bivalent antigen with bivalent antibody in
solution, the number of antibody molecules in the complex
being 1 (top row), 2 (middle row) or 5 (lowest row) - Reproduced from ref. 27.

is that, as shown in fig. 3, Scatchard plots can assume, for suitable values of S_o and of $K'(1)$, a downward concavity, similar to that which is usually interpreted as denoting positive cooperativity.

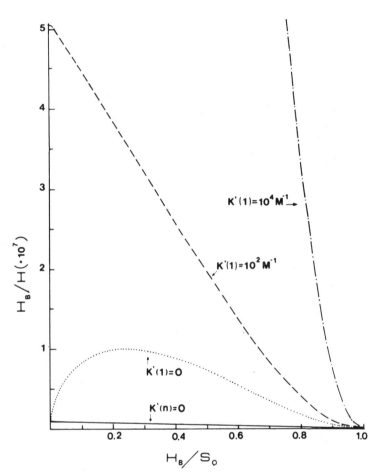

Fig. 3. Scatchard plot representation of the binding between bivalent antibody and bivalent antigen. H_B and H are the concentrations of bound and free antigenic determinants, respectively. S_o is the total concentration (assumed to be 1 uM) of antibody sites. The intrinsic dissociation constant of interaction between antigenic determinants and antibody sites was assumed to be 1 uM. In the lowest ($K'(n)=0$) trace, no ring formation was allowed. In the top one (at the extreme right), any ring could be formed, the value of $\overline{K}'$ being 10^4 (see text). In the other curves, this same assumption was maintained for rings with n 2, while monogamous rings (n=1) were excluded ($K'(1)=0$) or taken as less probable ($K'(1) = 10^2$). Unpublished data by Gandolfi and Strom.

Fig. 4 illustrates how similar shapes can indeed be occasionally
found when the binding of multivalent ligands to membranes is measured.

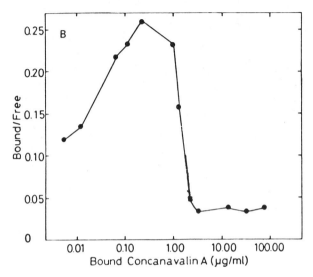

Fig. 4. Scatchard representation of concanavalin A binding to human
erythrocytes – Experimental data from ref. 30.

KINETIC CONTROL OF BINDING BY LATERAL DIFFUSION

The redistribution of membrane receptors, which is to be expected in the binding of a same multivalent effector to different receptors, can become the rate-limiting process if the lateral diffusion
within the plane of the membrane is slower than the binding itself.
This was demonstrated by studying how reduced and alkylated MOPC 315
myeloma protein, which is known to possess two independent sites
with a high affinity for dinitrophenol, binds to phosphatidylcholine
vesicles "doped" with known amounts (1%) of DNP-phosphatidylethanol-
amine. The binding was followed at $25^{o}C$ in a stopped flow rapid
mixing apparatus by monitoring the variations in tryptophan fluorescence of the MOPC protein upon binding to the DNP groups. Using
dimiristoylphospholipids, the kinetics could be analysed in terms
of a typical bimolecular reaction, with rate constants $k_{on} = 4 \times 10^{7}$
$M^{-1}s^{-1}$ and $k_{off} = 10$ s^{-1}. Using instead dipalmitoylphospholipids,
more complex kinetics were found, which, in a pseudo-first order
analysis, showed the presence of two consecutive processes: a bimolecular one, having a k_{on}^{appl} value approximately equal to that of
the previously mentioned k_{on}; and a monomolecular step, with

values of k_{app2} = 88 s^{-1} and k_{app-2} = 9 s^{-1} (fig. 5). It appears therefore that a higher viscosity of the lipid phase can result in a
slower rate of lateral diffusion of the vesicle-associated DNP
moieties, which can be reflected on a limitation in their availability for the binding to the second group of the bivalent MOPC protein.

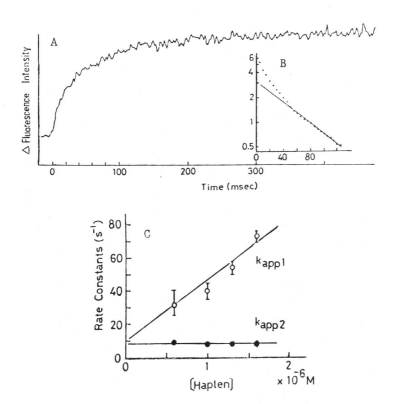

Fig. 5. Kinetics of the binding of MOPC 315 protein to DNP-doped phospholipid vesicles. In trace A is shown the time variation of
tryptophan fluorescence when a 26 nM solution of MOPC protein
is mixed, at pH 7.0 and 25°C, with a 75 µg/ml suspension of
dipalmitoylphosphatidylcholine vesicles containing 1% DNP-
dipalmitoylphosphatidylethanolamine (final DNP concentration:
1 µM). Trace B is a semi-logarithmic plot of the same data.
In trace C the values of the rate constants, k_{app1} and k_{app2},
are plotted against the hapten concentration.

CONCLUSIONS

The binding of multivalent effectors to membrane receptors can exhibit some peculiar characteristics deriving either from the possible formation of closed circular complexes or from the presence of rate-limiting step(s) in the translational diffusion of single receptors. Elucidation of these aspects in well-defined experimental conditions appears as a prerequisite to any speculation on the possible relations existing between effector binding to membrane receptors and cell stimulation.

ACKNOWLEDGEMENTS

The financial support of the Fondazione Cenci Bolognetti and of the Italian National Research Council is gratefully acknowledged.

REFERENCES

1. J.Schlessinger, Receptor aggregation as a mechanism for transmembrane signalling: models for hormone action, Dev.Cell Biol. 4:89-118 (1978).
2. C.DeLisi, The biophysics of ligand-receptor interactions, Quart. Rev.Biophysics 13:201-232 (1980).
3. R.P.Siraganian, W.A.Hook and B.B.Levine, Specific in vitro histamine release from basophils by bivalent haptens: evidence for activation by simple bridging of membrane bound antibody, Immunochem. 12:149-157 (1975).
4. D.M.Segal, J.D.Taurog and H.Metzger, Dimeric immunoglobulin E serves as a unit signal for mast cell degranulation, Proc.Nat. Acad.Sci.U.S.A. 74:2993-2997 (1977).
5. T.Ishizaka and K.Ishizaka, Triggering of histamine release from rat mast cells by divalent antibodies directed against IgE-receptor, J.Immunol. 120:800-805 (1978).
6. M.Dembo, B.Goldstein, A.K.Sobotka and L.M.Lichtenstein, Histamine release due to bivalent penicilloyl haptens: control by the number of cross-linked IgE antibodies on the basophil plasma membrane. J.Immunol. 121:354-358 (1978).
7. H.Krakauer, J.S.Peacock, B.G.Archer and T.Krakauer, The interaction of surface immunoglobulins of lymphocytes with highly defined synthetic antigens, in "Physical Chemical Aspects of Cell Surface Events in Cellular Regulation", C.DeLisi and R.Blumenthal, eds., p.345-362, Elsevier/North Holland, New York (1979).
8. A.K.Sobotka, M.Dembo, B.Goldstein and L.M.Lichtenstein, Antigen-specific desensitization of human basophils, J.Immunol. 122: 511-517 (1979).
9. C.DeLisi and R.Siraganian, Receptor crosslinking and histamine release.I.The quantitative dependence of basophil degranulation on the number of receptor doublets, J.Immunol. 122:2286-2292 (1979).

10. B.Schechter, H.Lis, R.Lotan, A.Novogrodsky and N.Sharon, The
    requirement for tetravalency of soybean agglutinin for induction
    of mitogenic stimulation of lymphocytes, Eur.J.Immunol. 6:145-
    149 (1976).
11. M.Feldmann, Induction of immunity and tolerance in vitro by
    hapten-protein conjugates. The relationship between the degree
    of hapten conjugation and the immunogenicity of DNP-Pol., J.
    Exp.Med. 135:735-753 (1972).
12. N.R.Klinman, The mechanism of antigenic stimulation of primary
    and secondary clonal precursor cells. J.Exp.Med.136:241-260
    (1972).
13. G.I.Bell, B lymphocyte activation and lattice formation,
    Transplant.Rev. 23:23-46 (1975).
14. M.Feldmann, J.G.Howard and C.Desaymard, Role of antigen struc-
    ture in the discrimination between tolerance and immunity by
    B cells, Transplant.Rev. 23:78-97 (1975).
15. G.G.B.Klaus and J.H.Humphrey, Mechanism of B cell triggering:
    studies with T-cell independent antigens, Transplant.Rev.
    23:105-118 (1975).
16. H.Waldmann and A.Munro, B cell activation, Transplant.Rev. 23:
    213-222 (1975).
17. F.M.Griffin Jr. and S.C.Silverstein, Segmental response of the
    macrophage plasma membrane to a phagocytic stimulus, J.Exp.Med.
    139:323-336 (1974).
18. H.M.Dintzis, R.Z.Dintzis and B.Vogelstein, Molecular determinants
    of immunogenicity: the immunon model of immune response. Proc.
    Nat.Acad.Sci.U.S.A. 73:3671-3675 (1976).
19. B.Vogelstein, R.Z.Dintzis and H.M.Dintzis, Specific cellular
    stimulation in the primary immune response: a quantized model,
    Proc.Nat.Acad.Sci.U.S.A. 79:395-399 (1982).
20. A.Tominaga, K.Takatsu and T.Hamaoka, Acceptor site(s) for T cell-
    replacing factor (TRF) on B lymphocytes.II.Activation of B cells
    by cross-linkage or aggregation of the TRF acceptor molecule,
    J.Immunol. 128:2581-2585 (1982).
21. J.M.Teale and N.R.Klinman, Tolerance as an active process,
    Nature 288:385-387 (1980).
22. A.C.E.King and P.Cuatrecasas, Peptide hormone-induced receptor
    mobility, aggregation and internalization, New England J.Med.
    305:77-88 (1981).
23. C.R.Kahn, K.L.Baird, D.B.Barrett and J.S.Flier, Direct demon-
    stration that receptor crosslinking or aggregation is important
    in insulin action, Proc.Nat.Acad.Sci.U.S.A. 75:4209-4213 (1978).
24. D.Baldwin Jr., S.Terris and D.F.Steiner, Characterization of
    insulin-like actions of anti-insulin receptors antibodies.
    Effects on insulin binding, insulin degradation, and glycogen
    synthesis in isolated rat hepatocytes, J.Biol.Chem. 255:4028-
    4034 (1980).
25. A.Gandolfi, M.A.Giovenco and R.Strom, Reversible binding of
    multivalent antigen in the control of B lymphocyte activation,
    J.theor.Biol. 74:513-521 (1978).

26. A.Gandolfi, M.A.Giovenco and R.Strom, Control of B lymphocyte
    activation through reversible binding of multivalent antigen:
    a simple model, in "Systems Theory in Immunology", C.Bruni,
    G.Doria, G.Koch and R.Strom,eds., p.37-51, Lecture Notes in
    Biomathematics, Springer-Verlag, Berlin (1979).
27. A.S.Perelson and C.DeLisi, Receptor clustering on a cell surface.
    I.Theory of receptor cross-linking by ligands bearing two
    chemically identical functional groups, Math.Biosciences 48:
    71-110 (1980).
28. A.Gandolfi and R.Strom, 'Avidity' plots as a tool for the eval-
    uation of antibody affinities toward multivalent antigens,
    in "Mathematical Modeling in Immunology and Medicine", G.I.
    Marchuk and L.N.Belykh, eds., p.237-246, North Holland, Amster-
    dam (1983).
29. H.Jacobson and W.H.Stockmayer, Intramolecular reaction in poly-
    condensations.I.The theory of linear systems,J.Chem.Phys. 18:
    1600-1606 (1950).
30. Y.Okada, A study of concanavalin A binding to human erythrocytes,
    Biochim.Biophys.Acta 648:120-128 (1981).

THE PROPERTIES OF COATED VESICLES AND THEIR ROLE IN RECEPTOR-MEDIATED ENDOCYTOSIS

H. Edelhoch and P. K. Nandi

Clinical Endocrinology Branch, NIADDK
National Institutes of Health
Bethesda, Md.

We would like to describe some molecular and chemical properties of a cytoplasmic organelle i.e. coated vesicles, CVs, which are the vehicle by which certain substances are transported both across cellular membranes and between cytoplasmic organelles. A considerable number of metabolites (hormones, viruses, lipo- and glycoproteins, etc.) are bound in localized regions of the plasma membrane, called coated pits, by specific receptors[1-5]. These receptor-ligand complexes are internalized by their budding from the membrane in a process referred to as receptor-mediated endocytosis[6,7]. Coated pits and vesicles also play an important role in limiting the growth of the plasma membrane of motor nerve terminals when synaptic vesicles fuse with them by transferring membrane to cisternal membrane[8]. Coated vesicles also appear to play a role in secretion by transporting newly synthesized glycoproteins from the RER to the Golgi and then to the plasma membrane[7,9].

In the internalization of receptor-ligand complexes, CVs become smooth vesicles (endocytic) soon after they are formed from coated pits by losing their protein coats. The smooth vesicles then migrate to their target tissues, usually the lysosomes. Before fusion with the target organelle, the vesicle divides, separating receptor and ligand. The receptor is then returned to the plasma membrane, ready for another cycle. In the case of plasma lipoproteins and glycoproteins, a receptor goes through one cycle in a few minutes [9,10].

Pearse first purified CVs in 1975[11] and identified clathrin as the protein responsible for the structure of the coat. The coat

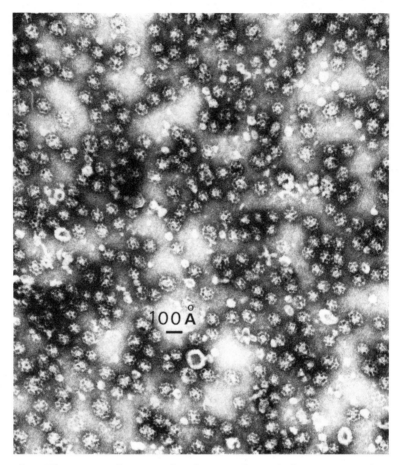

Fig. 1. Electron micrograph of coated vesicles prepared in 8% sucrose/^{2}H$_2$O step gradient (x 70,000).

was shown to have a very distinctive morphology in consisting of a network of pentagons and hexagons which form a porous, closed shell i.e. basket or cage[12,13]. By a process of quick-freeze and deep etching, Heuser was able to visualize by electron micrography the structure of clathrin in coated pits and the formation of CVs from coated pits[2]. Subsequently Unanue et al. [14] obtained the structure of clathrin by electron micrography showing that it had 3 long, bent arms which were joined at a central locus, i.e. a triskelion form. Cross-linking experiments indicated that a smaller protein, i.e. light chains, is present in native clathrin

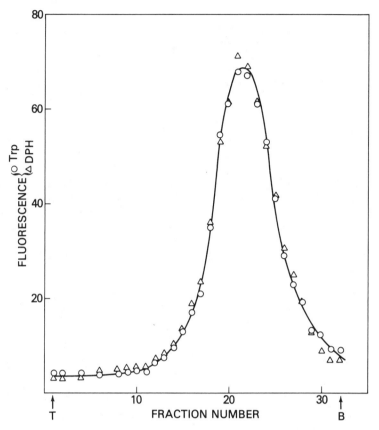

Fig. 2. Fluorescence analysis of coated vesicles after sucrose
gradient centrifugation (10-30%) at 27,000 rpm for 110
min at 20°C (o) Trp; (Δ) DPH. T is the top and B
is the bottom of the gradient.

and that it is stoichiometrically combined to the larger clathrin
chain[15,19].

   Clathrin is readily dissociated from CVs by raising the pH
from 6.5 to 7.5-8.5 or adding 2 M urea or Tris buffer[11,16-20].
This form of clathrin, considered as the native molecule, has
sedimentation constant near 8 and a molecular weight of about

610,000[21]. These values give a very large frictional coefficient, i.e. 3.0, which, however, is compatible with its very asymmetric structure. Clathrin therefore has the shape of a triskelion and consists of 3 long ($M_r \sim$ 180,000) and 3 short ($M_r \sim$ 35,000) polypeptide chains. Its form does not change with moderate variations in pH, temperature and ionic strength[21]. It readily self-associates to form coat structures at pH's below 7.0[18-21]. The sedimentation pattern of reassociated clathrin baskets shows 2 discrete size distributions with average rates of 150S and 300S[21]. Molecular weight of 25 and 100 x 10^6 were obtained by sedimentation equilibrium for the two types of baskets, respectively.[22]

Several interesting catalytic properties of CVs have been observed recently. CVs have been shown to contain an ATP-driven proton pump[23,24]. They also exhibit protein kinase activity in phosphorylating a 50,000 protein present in CVs[25]. The acidification of the interior of the vesicle is believed to be necessary for the dissociation of receptor and ligand in the endosome prior to the recycling of receptor to the membrane[26]. It has been demonstrated that CVs can fuse with lysosomes, the normal target organelle for most internalized ligands[27]. However, uncoated vesicles, UVs, fuse at a much greater rate than CVs. Thus, the shedding of the coat proteins is probably a prerequisite to fusion.

CVs dissociate in high concentrations of sucrose ($\sim$ 50%) into coat proteins and UVs. By substituting D_2O for sucrose, CVs can be purified in density gradients to give a preparation which shows a unimodal distribution of particle sizes[28]. Moreover, when they are sedimented on a 10-30% sucrose gradient, where they are stable, the ratio of protein to phospholipid in each fraction is constant (Figs. 1, 2) indicating a constant chemical composition in all the vesicles[28].

The molecular parameters and mass distribution of preparations of CVs were determined by sedimentation and quasielastic light scattering measurements[29]. The distribution of sedimentation coefficients is shown in Fig. 3. The mean value (fit to a Schulz distribution) was 225S. A diffusion coefficient of 4.8 x 10^{-8} cm^2/sec was determined by light scattering. A partial specific volume of 0.765 was obtained from the dependence of the sedimentation coefficients on density in H_2O-D_2O mixtures. This value is not far from that of normal proteins. This is due to the high percentage of protein present in CVs which is accounted for largely by presence of the coat proteins. Substituting these values in the Svedberg equation, a weight average molecular weight of 49 x 10^6 was calculated. The mass distribution of CVs was also determined from sedimentation equilibrium experiments by assuming a Schulz distribution of particle sizes[29]. A radius of 100 nm was computed from the diffusion coefficient by assuming the Stokes-

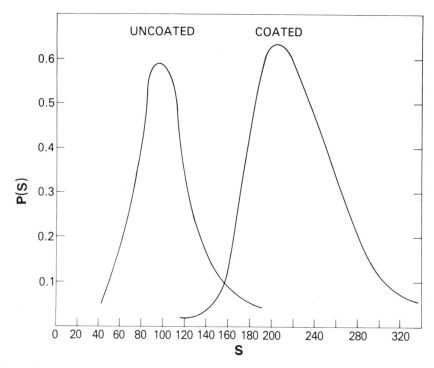

Fig. 3. Distribution of sedimentation coefficients P(s) of CVs and UVs.

Einstein relation ($D = kT/6r\eta$) for spheres. This value is some-what greater than that found by electron microscopy i.e. 75 nm. The unusual structure of the protein coat could modify the hydro-dynamic behavior of CVs and account for some of the difference in radii.

We have similarly characterized the properties of UVs. They contain a protein/phospholipid ratio of 60%/40%. The distribution of sedimentation coefficients is shown in Fig. 3. The mean value was 102S. A diffusion coefficient of 8.2×10^{-8} cm^2/sec was found by light scattering. When combined with a value of 0.83 for the partial specific volume, a weight average molecular weight of 13×10^6 was obtained. A radius of 59 nm was computed from the diffu-sion coefficient which is only slightly larger than the value of 52 nm observed by electron micrography.

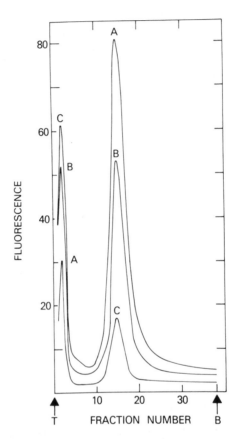

Fig. 4. Effect of concentration on the polymerization of clathrin at pH 6.40 in 0.10 M Mes. Sucrose gradient centrifugation was on a linear 10-30% gradient. Native clathrin concentrations were (A) 1.04, (B) 0.69 and (C) 0.40 mg/mL (before dialysis).

Clathrin can reassociate to form a coat or basket structure in the absence of membrane when the pH is reduced below about 7. The rate of reaction can be conveniently followed by either light scattering at 436 nm or absorbance at wavelengths above ~ 320 nm. Both the rate and equilibrium of basket formation increase rapidly with pH between pH ~ 7 and 5.8. The equilibrium appears to be only between clathrin monomers (8S) and baskets since intermediate size polymers were not observed when the reaction was followed by velocity ultracentrifugation[30] or by sucrose gradient analysis, Fig. 4[31]. In contrast to most self-associating systems, i.e. sickle cell hemoglobin[32], tubulin[33], etc., there does not appear to be a critical concentration for polymerization. In fact, the initial rates hardly vary with clathrin concentration. However, the rate of reaction, when followed with time, showed a high order, i.e. about 6[30]. It appears, therefore, that a very rapid growth process follows an initial, rate controlling, unimolecular activation process.

The rate and equilibrium of clathrin self-association are also dependent on the ionic strength of the solution. Increasing salt concentration decreases the rate and extent of basket formation. The effects of different salts, however, follow the Hofmeister series[30]. The latter normally reflects hydrophobic interactions whereas the former reflects electrostatic interactions. Both types seem to be necessary for clathrin polymerization. The strong pH dependence suggests either that certain groups need to be protonated (i.e. histidyl or carboxyl) or a pH-dependent conformational change is part of the activation step. It is also possible that the rate and equilibrium may be controlled, in part, by the net charge since increasing the pH above 5.8 also increases the negative electrical charge on clathrin.

We have also studied the formation of coat structure in the presence of UVs, i.e. the equilibrium between clathrin and UVs to form CVs. The extent of formation of protein and phospholipid components resulting from the dissociation of CVs at pH 8.5 can be analyzed after their separation on a 10-30% sucrose gradient[34]. The slower sedimenting clathrin boundary contains very little phospholipid whereas the phospholipid boundary (i.e. UVs) includes about 10-15% of the protein (Fig. 5). Electrophoresis in SDS of UVs reveals considerable amounts of protein in the 50-60,000 and 100-110,000 regions of the gel. The protein peak consists largely of clathrin and light chains, but includes small amounts of 100-110,000 proteins. [A protein of molecular weight about 110,000 has been separated from 8S clathrin on lysine-sepharose columns which appears to be necessary to form the 150S species. In the absence of this protein the 300S species is favored[31].] When the pH is returned to 6.5 (Fig. 5), both peaks largely disappear and form a boundary corresponding to CVs. At pH 7.5, the dissociation is less

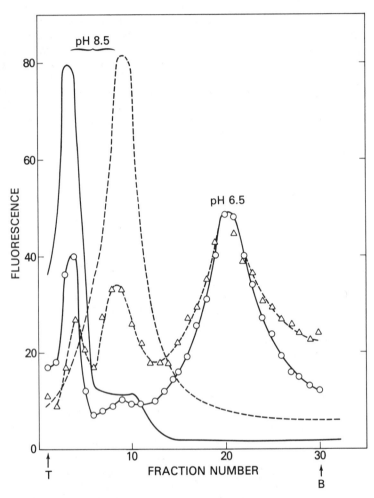

Fig. 5. Sucrose gradient centrifugation (10-30%) of coated ves-
         icles dissociated at pH 8.5, 0.10 M Tris (curves without
         symbols) and reassociation of dissociated coated vesicles
         at pH 6.5, 0.10 M Mes (curves with symbols). (○) Trp fluor-
         escence; (△) DPH fluorescence.

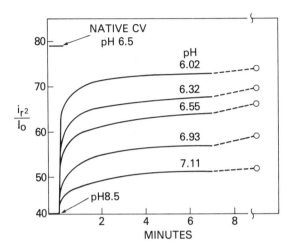

Fig. 6. Changes in light scatter with pH. Native coated vesicles
(0.25 mg/mL) at pH 6.5, 0.01 M Mes were brought to pH 8.5
by addition of 50 μL of 0.20 M Tris, pH 8.55, to 1 mL of
solution. The light scatter fell from 78 to 40 units.
This fall is a measure of the degree of dissociation of
coated vesicles. Solutions were then brought to lower pH
values. The rates of increase in light scatter were measured continuously for 7 min and then after 12 h (o). The
temperature was 23°C. The increase in light scatter is
a measure of the degree of reassociation of uncoated vesicles and clathrin.

complete than at pH 8.5; however, the reassociation is more reversible[34]. In accord with the ease of reformation of CVs, Unanue
et al. have reported an affinity constant of 10^9 L/mole for the
binding of clathrin to UVs[35].

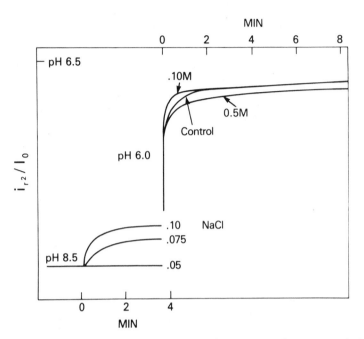

Fig. 7. Effect of NaCl on the rate and extent of reassociation.
The pH of native coated vesicles was increased from 6.5
to 8.5 with Tris with a 50% loss in light scatter. NaCl
was added at pH 8.5. A small increase in light scatter
occurred which was finished in 3-4 min. After 4 min, the
pH was reduced to 6.0 with Mes. A much faster and larger
increase in light scatter occurred. At pH 6.0, most of
the initial light scatter was recovered.

Since the self-association of clathrin and its reassociation
with UVs to form CVs are equilibrium reactions, we have compared
the two in order to assess the influence of the membrane on the
kinetics of coat formation. Kinetic observations by light scatter
indicated a rapid rate of reassociation between pH about 7 and 6
without much dependence on pH (Fig. 6). The extent of reassociation, however, varied with pH, in a manner similar to that of

clathrin self-association. A marked difference in rates was seen
in the effect of salt since the reassociation reaction was essen-
tially independent of NaCl between 0.10 and 0.50 M (Fig. 7) whereas
the rate and extent of clathrin polymerization decreased with increasing NaCl concentration and was strongly inhibited by 0.10 M
NaCl (Fig. 8). These differences in properties suggest that a protein is present in UVs which interacts with clathrin and modifies
its self-association to provide a new behavior pattern. The pattern is only partially affected since the characteristic lack of
dependence of initial rates on clathrin concentration remained the
same for CV reassociation. The influence of a membrane protein in

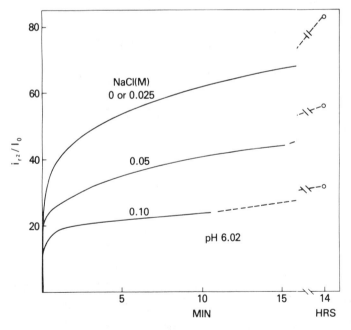

Fig. 8. Effect of NaCl concentrations on the rate of polymeriza-
         tion of clathrin at pH 6.33. Clathrin concentration was
         0.7 mg/mL in all solutions. The pH was adjusted with
         50 µl of 1 M MES to a final concentration of 0.045 M.

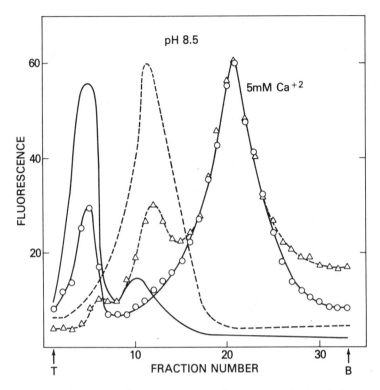

Fig. 9. Dissociation of coated vesicles at pH 8.5, 0.01 M Tris
(curves without symbols) and reassociation by 5 mM Ca²⁺
at pH 8.5, 0.01 M Tris (curves with symbols). (o) Trp
fluoroescence; (Δ) DPH fluorescence. Centrifugation on
a sucrose gradient (10-30%).

modifying the properties of clathrin coats in CVs was demonstrated
directly by Unanue et al. by digesting UVs with elastase[35]. After
this treatment, clathrin no longer combined with UVs.

Although univalent ions inhibit clathrin self-association and
are without much influence on clathrin reassociation with UVs, certain divalent ions have been shown to enhance basket formation and
inhibit CV dissociation[21,36]. We have measured the rates of self-
association of clathrin and found that the initial rates increase

linearly with Ca^{+2} concentration. Mn^{+2} has about 4-fold greater effect whereas Mg^{+2} had only a borderline influence on the rate[36].

The stimulation of clathrin self-association by Ca^{+2} was even more profound when observed by reassociation with UVs, since the reaction could now be made to occur at pH values much higher than was possible without Ca^{+2}. The reassociation occurred at pH 8.5 in the presence of 1-5 mM Ca^{+2}. Sucrose gradient analysis showed the presence of CVs sedimenting at normal rates which were homogeneous with respect to protein/phospholipid composition (Fig. 9). It appears that the binding of Ca^{+2} can substitute for protons in providing the appropriate form of the site or charge needed to initiate the polymerization of clathrin to form coat structure.

Various phenothiazine and antimalarials have been reported to inhibit receptor-mediated endocytosis by interfering with their membrane receptors[37-39]. Since clathrin is important to coated pit and coated vesicle formation, we have evaluated the effect of these drugs on its ability to form coat structure. These drugs bind to clathrin and accelerate the rate of its self-association to form baskets, i.e. coat structure. The phenothiazines, chloropromazine and trifluoperazine, were more effective than the antimalarials, i.e. quinacrine and chloroquine in enhancing clathrin polymerization[40]. All of these drugs were also effective in reassociating clathrin with UVs after dissociating CVs. (Unpublished observations of A. DiCerbo).

The mechanism of action of these drugs in inhibiting internalization of ligand-receptor complexes is not known. One possibility is that they combine with clathrin in coated pits, change their conformation, and thereby inhibit their budding from the membrane. A second is that they polymerize clathrin in the cytoplasm to form empty coats (baskets) which are then immobilized and therefore not available for coated pit formation. It is now known that the round trip of receptor from the plasma membrane into the cytoplasm and back to the membrane is several minutes for two receptors[5-10]. Consequently, the sequestration of even small amounts of any component involved in this cycle could have a strong inhibitory effect on the entire process.

REFERENCES

1. M. S. Brown and J. L. Goldstein, Proc. Natl. Acad. Sci. USA 76:3330 (1979).
2. J. E. Heuser, J. Cell Biol. 84:560 (1980).
3. B. M. F. Pearse and M. Bretscher, Ann. Rev. Biochem. 50:85 (1981).
4. I. H. Pastan and M. Willingham, Science 214:504 (1982).

5. S. M. Brown, R. G. W. Anderson, and J. L. Goldstein, Cell 32: 663 (1983).
6. J. L. Goldstein, R. G. W. Anderson, and M. S. Brown, Nature (London) 279:679 (1979).
7. M. G. Farquhar, Fed. Proc. 42:2407 (1983).
8. J. E. Heuser and T. S. Reese, J. Cell Biol. 57:315 (1973).
9. J. E. Rothman and R. D. Fine, Proc. Natl. Acad. Sci. USA 77: 780 (1980).
10. A. Dautry-Varsat, A. Ciechaover, and H. F. Lodish, Proc. Natl. Acad. Sci. USA 80:2258 (1983).
11. B. M. F. Pearse, J. Mol. Biol. 97:93 (1975).
12. R. A. Crowther, J. T. Finch, and B. M. F. Pearse, J. Mol. Biol. 103:785 (1976).
13. R. A. Crowther and B. M. F. Pearse, J. Cell Biol. 91:790 (1981).
14. E. Ungewickell and D. Branton, Nature 289:420 (1981).
15. T. Kirchhausen and S. C. Harrison, Cell 23:755 (1981).
16. A. L. Blitz, R. E. Fine, and P. A. Tosselli, J. Cell Biol. 75:135 (1977).
17. M. P. Woodward and T. F. Roth, Proc. Natl. Acad. Sci. USA 75: 4394 (1978).
18. W. Schook, S. Puszkin, W. Bloom, C. Ores, and S. Kochwa, Proc. Natl. Acad. Sci. USA 76:116 (1979).
19. B. M. F. Pearse, Proc. Natl. Acad. Sci. USA 73:1255 (1976).
20. J. H. Keen, M. C. Willingham, and I. H. Pastan, Cell (Cambridge, Mass.) 16:303 (1979).
21. H. T. Pretorius, P. K. Nandi, R. E. Lippoldt, M. L. Johnson, J. H. Keen, I. Pastan, and H. Edelhoch, Biochemistry 20: 2777 (1981).
22. P. K. Nandi, H. T. Pretorius, R. E. Lippoldt, M. L. Johnson, and H. Edelhoch, Biochemistry 19:5917 (1980).
23. D. K. Stone, X-S. Xie, and E. Racker, J. Biol. Chem. 258: 4059 (1983).
24. M. Forgac, L. Cantley, B. Wiedenmann, L. Altstiel, and D. Branton, Proc. Natl. Acad. Sci. USA 8:1300 (1983).
25. A. Pauloin, I. Bernier, and P. Jolles, Nature 298:574 (1982).
26. B. Tycko and F. R. Maxfield, Cell 28:643 (1982).
27. L. Altstiel and D. Branton, Cell 32:921 (1983).
28. P. K. Nandi, G. Irace, P. P. Van Jaarsveld, R. E. Lippoldt, and H. Edelhoch, Proc. Natl. Acad. Sci. USA 79:5881 (1982).
29. R. Nossal, G. H. Weiss, P. K. Nandi, R. E. Lippoldt, and H. Edelhoch, Arch. Biochem. Biophys. (1983 in press).
30. P. P. Van Jaarsveld, P. K. Nandi, R. E. Lippoldt, H. Saroff, and H. Edelhoch, Biochemistry 20:4129 (1981).
31. G. Irace, R. E. Lippoldt, H. Edelhoch, and P. K. Nandi, Biochemistry 21:5764 (1982).
32. J. Hofrichter, P. D. Ross, and W. A. Eaton, Proc. Natl. Acad. Sci. USA 71:4864 (1974).
33. F. Gaskin, C. R. Cantor, and M. L. Shelanski, J. Mol. Biol. 89:737 (1974).

34. P. K. Nandi, K. Prasad, R. E. Lippoldt, A. Alfsen, and H. Edelhoch, Biochemistry 21:6434 (1982).
35. E. R. Unanue, E. Ungewickell, and D. Branton, Cell 26:439 (1981).
36. P. K. Nandi, P. P. Van Jaarsveld, R. E. Lippoldt, and H. Edelhoch, Biochemistry 20:6706 (1981).
37. H. T. Haigler, M. C. Willingham, and I. Pastan, Biochem. Biophys. Res. Commun. 94:630 (1980).
38. D. Fitzgerald, R. E. Morris, and C. B. Saelinger, Cell 21: 867 (1980).
39. J. L. Salisbury, J. S. Condeelis, and P. Satir, J. Cell Biol. 87:132 (1982).
40. P. P. Van Jaarsveld, R. E. Lippoldt, P. K. Nandi, and H. Edelhoch, Biochem. Pharmacol. 31:793 (1982).

INTERACTION OF INSULIN WITH PLASMA MEMBRANE OF TARGET CELLS:

THE ROLE OF MEMBRANE FLUIDITY

Paolo Luly

Department of Cellular and Developmental Biology
University of Rome 'La Sapienza' and Department of
Biology, 2nd University of Rome, 'Tor Vergata'
Rome, Italy

INTRODUCTION

It is now well accepted that all of insulin's actions are the
result of its interaction with a specific receptor on the plasma
membrane of target cells. The concept of a membrane effect of the
hormone together with the suggestion of the presence of a
specific receptor dates from the early 50s when experimental
evidences, indicating both the binding to a target tissue (Stadie
et al., 1949) and a significant role in sugar trasport (Levine et
al., 1949), were published.

But in the mid 70s, when the role of cyclic amp as a second
messenger was well assessed in most cellular types investigated,
the problem arose whether insulin really needed such an
intracellular messenger (Goldfine, 1977). Proofs were available
both favouring this hypothesis and at the same time also
suggesting alternate pathways of intracellular mediation, but the
primary role of hormone-plasma membrane interaction was to be
considered as a basic concept (Pilkis and Park, 1974). In the
same period, the recently established hypothesis of the plasma
membrane as a 'fluid mosaic' (Singer and Nicolson, 1972)
stimulated provocative suggestions about the mechanism of action
of peptide hormones, which was considered to be strictly
dependent on the fluidity conditions of membrane microenvironment

39

as far as this structure modulates the interactions of a
'hormone-receptor' complex with an effector enzymatic molecule
such as adenylate cyclase (Cuatrecasas, 1974) or cyclic
AMP-phosphodiesterase (Loten and Sneyd, 1970; House et al., 1972;
Tria et al., 1977).

The first suggestion that the mechanism of the action of
insulin could in some way be mediated by changes of membrane
'lipid fluidity' (*) came from Farias' studies (Farias et al.,
1975; Farias, 1980). This Argentine investigator demonstrated
that, as in the behaviour of plasma membrane-bound cooperative
enzymes, insulin decreases membrane fluidity both in E. coli and
in mammalian erythrocytes. Contemporarily, possible changes of
membrane configuration were suggested as a consequence of
insulin-receptor interactions as the very first step of hormonal
action by Hepp (1977).

In an attempt to rationalize the manifold actions of insulin
on target cells, Goldfine (1978) has suggested a kind of
classification on the basis of the temporal sequence of these
actions. Thus, according to Goldfine's proposal, insulin effects
at various subcellular levels can be classified as: i) rapid,
restricted to cell plasma membrane; ii) intermediate, concerning
cytosol, E.R., ribosomes, mitochondria, lysosomes (basically
modulation of enzymatic activities); iii) delayed, at nuclear
level modulating DNA and RNA synthesis.

The aim of this presentation is that to convey your
interest on an array of experimental observations which, although
restricting our attention to rapid insulin actions, point to a
relationship between the mode of action of the hormone and the
physical state of target cell plasma membrane.

----

* The term 'lipid fluidity' is used as a general term (Brasitus
and Schachter, 1980; Schachter et al., 1982) to express the
relative motional of freedom of the bilayer lipid molecules in
its broadest meaning: including different types of motion (i.e.
rotational or lateral) combining both rate and extent of
movement. The appreciation of 'lipid fluidity' in such a general
sense can be achieved by steady-state fluorescence polarization
of lipid soluble probes which allows the estimation of anisotropy
parameters reflecting the overall motional freedom of the probe.

THE ROLE OF LIPIDS IN INSULIN-PLASMA MEMBRANE INTERACTIONS

It is well established that the insulin receptor is a glycoproteic heterotetramer with a molecular weight of about 360,000 daltons (Cuatrecasas, 1973; Jacobs and Cuatrecasas, 1981). Lipids were not considered particularly relevant to the binding process until very recently.

Insulin binding

It has been shown by Ginsberg and coworkers (1981) that Friend erythroleukemia cells increase their binding capacity and decrease their affinity for the binding of insulin when the bulk membrane fluidity is increased after culturing the cells in vitro in media containing supplemental unsaturated fatty acids; these investigators obtained the same qualitative results in vivo, growing Ehrlich ascites cells in mice fed with a diet particularly rich of unsaturated fatty acids (Ginsberg et al., 1982). Strictly comparable observations were reported, by Gould and coworkers (1982), on insulin receptor isolated from turkey erythrocytes and reconstituted into vesicles of controlled lipid composition: the increase of phospholipid saturation increases significantly binding affinity and decreases accordingly the number of available binding sites. Finally, the increase of lipid saturation by appropriate in vivo fat feeding decreases binding capacity of isolated rat liver plasma membrane leaving binding affinity basically unaffected (Sun et al., 1977).

The interpretation of this body of results, offered by Ginsberg and coworkers, was that the decrease of membrane lipid fluidity increases a tendency to self-association of the polymeric receptor thus presenting its high affinity binding site but conversely reducing its binding capacity; the reverse will take place under conditions of increasing membrane lipid fluidity. It is worth mentioning in this connection that the role of membrane fluidity has not been considered so far in the light of the 'receptor-mediated internalization' hypothesis (Steiner, 1977; Kahn, 1979), even if a clustering process for receptors is considered as obligatory for endocytosis to occurr, thus requiring the diffusion of receptor molecules in the plane od the membrane and, of course, optimal fluidity conditions (Singer, 1976). It appears, therefore, that the quality of the surrounding lipid environment can alter the properties of the insulin receptor, thus suggesting that localized changes of plasma membrane lipid fluidity could in some way affect the interaction of insulin with target cells.

Plasma membrane-bound enzymes and transport phenomena

At variance with the mechanism of action of glucagon, for which the coupling between receptor and adenylate cuclase, as well as the cyclase itself, are modulated by membrane lipid fluidity (Levitzky and Helmreich, 1979; Houslay, 1981), no such a clear relationship exists between insulin binding to target cells and membrane-linked processes in the general framework of the modulation of plasma membrane lipid fluidity. Only glucose transport and ion pumping activity have been so far related to membrane fluidity with reference to their insulin sensitivity as we shall see in some detail.

As partly reported before, Farias and coworkers (Massa et al., 1975; Moreno and Farias, 1976) first suggested that a decrease of membrane fluidity could be a general phenomenon in the mechanism of action of insulin: their proposal followed experimental reports in which such an insulin effect could be inferred from alterations of the cooperative behaviour of (Na-K)-ATPase and acetylcholinesterase in rat erythrocyte as well as of Ca-ATPase of E.coli membranes. These authors confirmed later this observation after in vivo appropriate feeding coupled with insulin i.p. addmistration which induced, in rat erythrocytes, the same response of acetylcholinesterase cooperative behaviour: a phenomenon to be considered strictly dependent on a decreased membrane fluidity (Unates and Farias, 1979). Results very similar to those reported by Farias were obtained on our laboratory on isolated rat liver plasma membranes, where the cooperative behaviour of (Na-K)-ATPase is reduced by an insulin in vitro treatment which can be related to a decrease of membrane lipid fluidity (Baldini et al., 1979).

It is worth mentioning in this context that a very recent and extended review of insulin effects on ion transport (Moore,

1983), in an attempt to give reasonable explanations of the
mechanism of hormone action upon membrane transport, while
excluding direct effects of insulin on the ion pump as well as
any role for cyclic nucleotides and stressing the contradictory
evidences of a calcium involvement, gives good experimental
support to a mediation of insulin action through induced changes
in the physical state of the plasma membrane.

The role of membrane fluidity in the regulation of hexose
transport, as related to insulin sensitivity, has been
investigatd in fat cells. Amatruda and Finch (1979), working on
adipocytes, reported that the physical state of membrane lipid
can modify hexose uptake ad well as the action of insulin on
these cells. These investigators suggested that, at 37°C, the
effect on membrane fluidity is a hormonally mediated one, as
assessed by thermal transition curves for glucose uptake in which
the temperature transition is detectable only in the presence of
hormone; the explanation suggested is that the presence of
separate lipid domains migh account for the differential hormonal
sensitivity of the transport system.

Contemporarily, Melchior and Czech (1979) reported that the
operativity of a reconstituted glucose transport system
(liposomes obtained from phospholipids extracted from adipocyte
plasma membranes in which were reinserted membrane proteins
previously removed) was optimal with a fully fluid lipid bilayer.
These authors suggested that the insulin stimulation of glucose
transport might involve changes of the physical properties of
lipids surrounding the transport system; such a result could be
accomplished by hormone through de novo synthesis of membrane
lipids as well as by the modification of the existing bilayer.
Basically, Czech supported experimentally (Pilch et al., 1980)
the hypothesis that hexose transport activation by insulin, in
adipocytes, might involve an increase of membrane lipid fluidity:
basal glucose transport was unaffected by decreased fluidity and
insulin-activated transport is unaffected by increased fluidity
but inhibited by its decrease. Two possible explanation were
offered for such a mechanism of action of insulin: i) an
insulin-dependent increase of bilayer fluidity in restricted
membrane regions in which the carrier system resides; ii) insulin
increases the partition coefficient of the hexose carrier for
more fluid regions of adipocyte plasma membrane. Czech and
co-workers (Pilch et al., 1980) point to a generalized fluidizing

effect of insulin on adipocyte plasma membrane, such an effect is
rationalized in relation to the increased transport capacity of
the hexose transport system: namely these investigators believe
that the insulin stimulatory effect is not due to an increased
number of transport sites, but to an increased activity of a
fixed number of transport sites which, in turn, could be
dependent on altered fluidity condition. Recent observations on
mice 3T3 fibroblasts (Yuli et al., 1982) support this latter
interpretation of an increased activity of carrier sites possibly
dependent, immediately after hormone binding, on a decreased
transport activation energy linked to optimal fluidity
conditions.

DIRECT EFFECTS OF INSULIN ON PLASMA MEMBRANE MICROENVIRONMENT

   One could wonder, at this stage, whether there is any
evidence of a diret effect of insulin on plasma membrane fluidity
of target cells. Indeed, few observations have been published so
far on this topic, on which I will briefly report in the
following.

   Bailey and coworkers (1978) reported, on adipocyte plasma
membrane, an increasing lipid fluidity after insulin in vitro
trreatment, using fluorescence polarization techniques. On the
other hand, such an observation could not be confirmed by ESR
studies, always carried out on adipocyte plasma membranes, by
independent investigators (Amatruda and Finch, 1979; Sauerheber
et al., 1980).

   Experiments carried out in vitro on isolated hepatocyte
plasma membrane with two different fluorescent methods, i.e. the
fluorescence polarization of 1,6-diphenyl-1,3,5-hexatriene (Luly
and Shinitzky, 1979) and the excited dimer formation with the
fluorescent probe pyrene (Luly et al., 1979) suggest that the
insulin negatively affects membrane lipid fluidity. In
particular, the effect of insulin on isolated hepatocyte plasma
membranes proved to be strictly specific for native insulin,
respondent to hormone concentrations down to 10^{-10}M and sensitive
to changes of cholesterol/phpspholipid ratio: namely the increase
of this ratio, indicating a decrease of membrane fluidity,
blocked any hormonal effect. The stiffening of rat liver plasma

membrane after an insulin in vitro treatment was very recently
confirmed by detailed studies of Schroeder (1982).

Finally, reports have appeared which indicate that insulin is
able to affect membrane fluidity of human erythrocyte ghosts. In
particular, a negative modulation of membrane fluidity, with a
dose-dependence of the phenomenon in the $10^{-10}\underline{-}10^{-8}$M range of
hormone concentration was reported (Luly et al., 1981); but, in
contrast, membrane fluidity of erythrocyte ghosts from both
diabetic and healthy donors was also reported to be enhanced
after insulin incubation in vitro, using the pyrene 'excimer'
formation method (Bryszewska and Leyco, 1983). It is also worth
mentioning, in this respect, that insulin was recently reported
to affect surface charge and electrophoretic behaviour of both
human and avian erythrocytes (Coker et al., 1981; Chen et al.,
1981).

From this body of observations concerning direct insulin
effects on plasma membrane fluidity of target cells it is not
possible to draw unequivocal evidences. It has been suggested
that these disparate effects of the hormone on membrane fluidity
may be due to a probe-linked sampling of different regions of the
plasma membrane, depending on selective partition properties of
different probes employed (Pilch et al., 1980; Sauerheber et al.,
1980; Luly and Shinitzky, 1979). But, as a matter of fact, it is
becoming more and more difficult to rule out a direct action of
insulin on membrane microenvironment, which - by the way - was
reported for insulin also on synthetic phospholipid bilayers
(Kafka, 1974).

IS THERE ANY INVOLVEMENT OF PLASMA MEMBRANE LIPID FLUIDITY IN
INSULIN-LINKED DISEASES?

Few experimental reports are available, so far, indicating
plasma membrane alterations in diabetes. It has been shown that,
in streptozotocin-induced diabetes, sialic acid and cholesterol
content significantly decrease both in liver and in erythrocyte
plasma membranes (Chandramouli and Carter, 1975); in addition, a
decreased fatty acid unsaturation was reported in comparable
experimental conditions (Eck et al., 1979). On the other hand,
reduced human erythrocyte deformability was demonstrated in

diabetes (McMillan et al.,1978): a phenomenon which can be
reversed by an in vitro treatment with insulin at physiological
concentration (Juhan et al., 1981); it has be recalled, in this
connection, that reduced erythrocyte deformability is attributed
to an increased 'viscosity' - generally speaking - of cell
membrane (McMillan et al., 1978) which affects the rehological
properties of circulating cells.

More direct evidences showing changes of lipid fluidity in
plasma membranes of circulating cells from human diabetic
subjects have also been published. A decrease of membrane
fluidity was reported in lymphocytes using the fluorescence
polarization technique (Cheung et al., 1980). In contrast,
independent investigations carried out on erythrocytes from
diabetic patients point to a generalized decrease of membrane
fluidity as assessed by both a fluorescent tecnique (Baba et al.,
1979; Bryszewska and Leyko, 1983) and ESR methods (Kamada and
Otsuji, 1983).

CONCLUSIONS

There is an increasing interest for any possible involvement
of plasma membrane properties in the mechanim of action of
insulin, with special attention for the diabetic state (McMillan,
1983). Collecting data on this very new approach to the study of
insulin effect at cellular level does not give an unequivocal
answer but offers a variety of information, perhaps depending on
the relationship between a peculiar target tissue and the
hormonal molecule.

All reported evidences suggest that insulin affects, in some
way, the lipid environment of plasma membrane: a situation which
could be of some relevance for the physiological role of proteins
embedded in the bilayer. In this context, how the interaction of
the hormone with its receptor could influence membrane lipid as
well has how transient this effect might be is far from clear. On
the other hand, whether this possible effect on membrane lipids
must be considered in order to explain the rapid actions of
insulin or might also act as a 'signal' for the regulation of
intracellular events, is matter for speculation. Still, for all
those interested to the 'insulin puzzle', a new argument for

debate and active research has now definitely introduced.

ACKNOWLEDGEMENTS

    P.L. is the recipient of a grant from the Italian National
Research Council, Project "Preventive and Rehabilitative
Medicine, SubProject Mechanisms of Aging".

REFERENCES

Amatruda, J.M. and Finch, E.D., 1979, Modulation of hexose uptake
    and insulin action by cell membrane fluidity, J. Biol.
    Chem., 254:2619.
Baba, Y., Kai, M., Kamada, T., Setoyama, S. and Otsuji, S., 1979,
    Higher levels of erythrocyte membrane microviscosity in
    diabetes, Diabetes, 28:1138.
Bailey, I.A., Garratt, C.J. and Wallace, S.M., 1978, An effect of
    fluorescent probes and of insulin on the structure of
    adipocyte membranes, Biochem. Soc. Trans., 6:302.
Baldini, P., Incerpi, S. and Luly, P., 1979, unpublished results.
Brasitus, T.A. and Schachter, D., 1980, Lipid dynamics and lipid-
    protein interactions in rat enterocyte basolateral and
    microvillous membranes, Biochemistry, 19:2763.
Bryszewska, M. and Leyko, W., 1983, Effect of insulin on human
    erythrocyte membrane fluidity in diabetes mellitus,
    Diabetologia, 24:311.
Chandramouli, V. and Carter, J.R., 1975, Cell membrane changes in
    chronically diabetic rats, Diabetes, 24:257.
Chen, S.M., Wrigglesworth, J.M., Perry, M.C., and Plummer, D.T.,
    1981, Effect of insulin on the electophoretic mobility of
    chick and human erythrocytes, Biochem.Soc.Trans., 9:128.
Cheung, H.C., Almira, E.C., Kansal, P.C., and Reddy, W.J., 1980,
    A membrane abnormability in lymphocytes from diabetic
    subjects, Endocrine Res.Commun., 7:145.
Coker, E.N., Perry, M.C., and Plummer, D.T., 1981, Changes in the
    zeta potential of chick erythrocytes after the addition of
    insulin, Biochem.Soc.Trans., 9:89.
Cuatrecasas, P., 1973, Insulin receptor of liver and fat cell
    membranes, Federation Proc., 32:1838
Cuatrecasas, P., 1974, Insulin receptors, cell membranes and
    hormone action, Biochem.Pharmacol., 23:2353.

Czech, M.P., 1980, Insulin action and the regulation of hexose transport, Diabetes, 29:399.

Eck, M.G., Wynn, J.O., Carter, W.J., and Faas, F.H., 1979, Fatty acid desaturation in experimental diabetes mellitus, Diabetes, 28:479.

Farias, R.N., 1980, Membrane cooperative enzymes as a tool for the investigation of membrane structure and related phenomena, Adv.Lipid Res., 17:251.

Farias, R.N., Bloj, B., Morero, R.D., Sineriz, F., and Trucco, R.E., 1975, Regulation of allosteric membrane-bound enzymes through changes in membrane lipid composition, Biochem. Biophys.Acta, 415:231.

Ginsberg, B.H., Brown, T.J., Simon, I., and Spector, A.A., 1981, Effect of the membrane lipid environment on the properties of insulin receptors, Diabetes, 30:773.

Ginsberg, B.H., Jabour, J., and Spector, A.A., 1982, Effect of alterations in membrane lipid unsaturation on the properties of the insulin receptor of Ehrlich ascites cells, Biochim.Biophys. Acta, 690:157.

Goldfine, I.D., 1977, Does insulin need a second messenger? Diabetes, 26:148.

Goldfine, I.D., 1978, Insulin receptors and the site of action of insulin, Life Sci., 23:2639.

Gould, R.J., Ginsberg, B.H., and Spector, A.A., 1982, Lipid effects on the binding properties of a reconstituted insulin receptor, J.Biol.Chem., 257:477.

Hepp, K.D., 1977, Studies on the mechanism of insulin action: basic concepts and clinical implications, Diabetologia, 13:177.

House, P.D.R., Poulis, P., and Weidemann, M.J., 1972, Isolation of plasma membrane subfraction from rat liver containing an insulin-sensitive cyclic AMP phosphodiesterase, Eur.J.Biochem., 24:429.

Houslay, M.D., 1981, Mobile receptor and collision coupling mechanisms for the activation of adenylate cyclase by glucagon, in: 'Advances in Cyclic Nucleotide Research', vol.14, J.E. Dumont, P.Greengard, and G.A.Robison,eds., p.111, Raven Press, New York.

Jacobs, S., and Cuatrecasas, P., 1981, Insulin receptors: structure and function, Endocrine Rev., 2:25.1

Juhan, I., Vague, P., Buonocore, M., Moulin, J.P., Calas, M.F., Vialettes,, B., and Verot, J.J., 1981, Effects of insulin on erythrocyte deformability in diabetics - relationship between erythrocyte deformability and platelets aggregation,

Scand.J.Clin.Lab.Invest., 41(Suppl.):156.

Kafka, M.S., 1974, The effect of insulin on the permeability of phosphatidyl choline bimolecular membranes to glucose, J.Membr.Biol., 18:81.

Kahn, C.R., 1979, Open question: what is the molecular basis for the action of insulin, Trends Biochem.Sci., 4:263.

Kamada, T., Otsuji, S., 1983, Lower levels of erythrocyte membrane fluidity in diabetics patients, Diabetes, 32:585.

Levine, R., Goldstein, M.S., Klein, S., and Huddlestone, B., 1949, The action of insulin on the distribution of galactose in eviscerated nephrectomized dogs, J.Biol.Chem., 179:985.

Levitzky, A., and Helmreich, E.J.M., 1979, Hormone-receptor-adenylate cyclase interactions, FEBS Lett., 101:213.

Loten, E.G., and Sneyd, J.G.T., 1970, An effect of insulin on adipose tissue adenosine 3':5'-cyclic monophosphate phosphodiester- ase, Biochem.J., 120:187.

Luly, P., and Shinitzky, M., 1979, Gross structural changes in iso- lated liver cell plasma membranes upon binding of insulin, Biochemistry, 18:445.

Luly, P., Crifo', C., and Strom, R., 1979, Effect of insulin on lat- eral diffusion of pyrene in rat liver plasma membrane, Experientia, 35:1300.

Luly, P., Baldini, P., Incerpi, S., and Tria, E., 1981, Insulin ef- fect in vitro on human erythrocyte plasma membrane, Experien- tia, 37:431.

Massa, E.M., Morero, R.D., Bloj, B., and Farias, R.N., 1975, Hormone action and membrane fluidity: effect of insulin and cortisol on the Hill coefficients of rat erythrocyte membrane-bound acetylcholinesterase and (Na-K)-ATPase, Biochem.Biophys. Res.Commun., 66:115.

McMillan, D.E., Utterback, N.G., and La Puma, J., 1978, Reduced erythrocyte deformability in diabetes, Diabetes, 27:895.

McMillan, D.E., 1983, Insulin, diabetes and the cell membrane: an hypothesis, Diabetologia, 24:308.

Melchior, D.L., and Czech, M.P., 1979, Sensitivity of the adipocyte D-glucose transport system to membrane fluidity in reconsti- tuted vesicles, J.Biol.Chem., 254:8744.

Moore, R.D., 1983, Effects of insulin upon ion transport, Biochim. Biophys.Acta, 737:1.

Moreno, H., and Farias, R.N., 1976, Insulin decreases bacterial membrane fluidity. It is a general event in its action? Biochem.Biophys.Res.Commun., 72:74.

Pilch, P.F., Thompson, P.A., and Czech, M.P., 1980, Coordinate mod-

ulation of D-glucose transport activity and bilayer fluidity in plasma membranes derived from control and insulin-
treated adipocytes, Proc.Natl.Acad.Sci.USA, 77:915.

Pilkis, S.J., and Park, C.R., 1974, Mechanism of action of insulin,
Ann.Rev.Pharmacol., 14:365.

Sauerheber, R.D., Lewis, U.J., Esgate, J.A., and Gordon, L.M., 1980,
Effect of calcium, insulin and growth hormone on membrane
fluidity, Biochim.Biophys.Acta, 597:292.

Schachter, D., Cogan, U., and Abbot, R.E., 1982, Asimmetry of lipid
dynamics in human erythrocyte membranes studied with
permeant fluorophores, Biochemistry, 21:2146.

Schroeder, F., 1982, Hormonal effects on fatty acids binding and
physical properties of rat liver plasma membranes, J.Membrane
Biol., 68:1.

Singer, S.J., 1976, The fluid mosaic model of membrane structure:
some applications to ligand-receptor and cell-cell interactions, in: 'Surface Membrane Receptors', R.A.Bradshaw,
W.A.Frazier, R.C.Merrell, D.I.Gottlieb, and R.A.Hogue-Angeletti,
eds., p.1, Plenum Press, New York.

Singer, S.J., and Nicolson, G.L., 1972, The fluid mosaic model of
the structure of cell membranes, Science, 175:7820.

Stadie, W.C., Hargaard, N., Hills, A.G., and Marsh, J.H., 149, The
chemical combination of insulin with muscle (diaphragm) of
normal rat, Am.J.Med.Sci., 218:265.

Steiner, D.F., 1977, Insulin today, Diabetes, 26:322.

Sun, J.V., Tepperman, H.M., and Tepperman, J., 1977, A comparison
of insulin binding by liver plasma membranes of rat fed
a high glucose diet or high fat diet, J.Lipid Res., 18:533

Tria, E., Scapin, S., Cocco, C., and Luly, P., 1977, Insulin-sensitive adenosine 3',5'-cyclic monophosphate phosphodiesterase
of hepatocyte plasma membrane, Biochim.Biophys.Acta, 496:77.

Unates, L.E., and Farias, R.N., 1979, In vivo modulation of rat erythrocyte acetylcholinesterase by insulin in normal and diabetic conditions, Biochim.Biophys.Acta, 568:363.

Yuli, I., Incerpi, S., Luly, P., and Shinitzky, M., 1982, Insulin
stimulation of glucose and amino acid transport in mouse
fibroblasts with elevated membrane microviscosity, Experientia,
38:1114.

EVALUATION OF TARGET CELL RESPONSIVENESS TO PEPTIDE HORMONES: A POSSIBLE ROLE FOR Na/K ATPase

*Roberto Verna,**Pierre Braquet,***Javier Diez
and ***Ricardo Garay

Institute of General Pathology, University of Rome,
viale Regina Elena, 324 -00161 Roma, Italy (*)

I.H.B. Laboratories, Le Plessis Robinson, France (**)

INSERM U7, Hôpital Necker, 75015 Paris, France (***)

The essential components of an endocrine system can be outlined as follows:
1. Synthesis and release of hormone by the proper endocrine gland.
2. Delivery to target cells by transport through the body fluids.
3. Recognition by target cells, through the specific receptor.

A fundamental concept in endocrinology is that hormone levels rise and fall, often rapidly, in response to changing body needs. We are much less comfortable with the idea that normally the target cell also fluctuates continuously and often widely in its responsiveness to hormone. Thus a given concentration of hormone will have a large or small effect depending on the state of the target cell.

Therefore a further step, fundamental to completing the action of the hormone, should be added to those mentioned above:

4. Transmission of the message inside the target cell.

51

What are the tools available to evaluate the efficiency of this step? When a hormone acts on a target cell, it typically does not activate a single process but rather regulates numerous biochemical events within the cell. In some cases the achievement of its biological effect may be recognized back at the level of the secretory cell, resulting in a reduction in hormone secretion, but most hormone-mediated events at the target cell are not recognized outside it.

At the same time, laboratory techniques in endocrinology are able to evaluate the amount of hormone released by its endocrine gland as well as that transported by the blood, and to estimate the distribution and efficiency of receptors on the cell surface, but seem to be inadequate to detect the transmission of the endocrine message inside the cell. In other words, a certain hormone may be normally synthesized, released, transported, and received, thus giving every appearance of normal endocrine activity, but its message may be wrongly translated inside the cell or not translated at all , with the result of an endocrine inefficiency. We therefore need to be able to assay this sensitivity, if not directly, at least indirectly, through systems that are affected by the transmission of the hormonal message (and possibly only by this step) and are thus able to inform us about the occurrence of such a transmission.

As far as water-soluble hormones and some prostaglandins are concerned, we know that step 4 consists in most cases in the activation of adenylate cyclase which, by raising cAMP levels, activates cAMP-dependent protein kinases, which in turn are responsible for the phosphorylation of enzymes. Since cAMP and its dependent protein kinases have also been reported to play a regulatory role in cell differentiation and growth (Ryan and Heidrick, 1968; Johnson et al., 1972) and in transport processes particularly in the regulation of electrolyte balance through the cell membrane (Garay, 1982), our study has been directed towards evaluating the possibility of using ion transport and/or cell proliferation as parameters to determine the responsiveness of target cells to peptide hormones.

A first approach to this problem was made by Tria et al. (1974), when they demonstrated on the one hand that cAMP reduced the activity of Na/K ATPase and on the other that endogenous phosphorylations were inhibited by the cyclic nucleotide, whereas exogenous phosphorylations were stimulated. This parallel effect of cAMP on Na/K ATPase and on the phosphorylation of membrane proteins suggested a role of endogenous phosphorylations in the regulation of Na/K ATPase activity. Starting from this consideration, our effort has been developed along two lines.

1. ROLE OF CYCLIC AMP IN THE CONTROL OF ION TRANSPORT

In living cells, Na entry into cells tends to be compensated by Na extrusion to the extracellular compartment, coupled to some extent with the entry of extracellular K driven by an active Na/K pump (Glynn and Karlish, 1975; Garrahan and Garay, 1976).

Since any alteration in the biological disposition of sodium would be expected to involve also the Na/K ATPase system and essential hypertension appears to result from a constellation of genetic and environmental factors of which an excess of Na intake is the most important (Dahl, 1972), studies have been directed towards evaluating the role of the sodium pump in hypertension by comparing the enzymatic activity and age in various tissues, as well as assaying the cAMP and cGMP concentrations in all our samples (Verna and Muraro, 1983c). Evidence is brought that hypertensive status is a disorder whose effects on metabolism can be found in all tissues and are not localized in vessels only. On the basis of the results, however, some considerations are worthy of discussion.

1. Hypertension is a condition that increases with age. The behavior of blood pressure in spontaneously hypertensive rats (SHR) has an increasing mode with respect to age, while age has only a small influence on blood pressure in the controls, and this effect is restricted to the very early weeks of life.

2. Almost all kinds of tissue show significant involvement of the Na/K ATPase enzymatic complex during the hypertensive condition.

3. This involvement of Na/K ATPase is different in intensity if related to age and/or to different tissues. This means that the same tissue has different responses at different ages and/or that different tissues have, at the same age, a different

sensitivity to the hypertensive stimulus.

Some questions then arise: (a) Is the primary hypertension
due to an alteration of the sodium pump itself? If so, why does
this change have different quantitative expressions in the
different tissues? Or (b) is the alteration in the sodium pump a
consequence of a modified metabolic activity of the tissues
involved?

We showed modified Na/K ATPase activity in the various
tissues tested. If the defect were due to the pump itself it
should be present to the same extent, in all tissues; but the
various tissues are involved in different ways. Liver and
erythrocyte membranes are unaffected, both with respect to
controls and with increased age. Thus the only way convincingly
to relate this phenomenon to a primary modification of the sodium
pump would be to demonstrate the existence of a different ATPase
for each tissue that would result in a non-uniform expression.
For this reason, we explored the possibility that Mg ATPase is
involved, but found no significant differences between
hypertensive and healthy subjects.

Thus, we believe that changes in the sodium pump represent
a secondary expression of a hypertension-induced metabolic
change. We prefer to hypothesize an alteration in the CNS
affecting both peripheral tone and hormonal homeostasis. This
hypothesis is supported by the finding of a large anomaly not
only in the cAMP content of peripheral tissues but also in the
cAMP/cGMP ratio which in turn would suggest that in addition to
the phosphodiesterases (PDE) (D'Armiento et al., 1980) other
regulatory systems for cAMP levels may be involved. Finally, the
behavior of both Na/K ATPase and cyclic nucleotide content in the
cerebral cortex indicates an altered metabolism of the nerve
cells in essential hypertension. Such an alteration could be
convincingly responsible for a neuroendocrine disequilibrium
involving the peripheral content of cyclic nucleotides.

These conclusions agree with previous findings (Suarez et
al., 1980) of an increase in blood pressure produced by
hypothalamic lesions and associated with significant hemodynamic
alterations. In this case we support the observation by Garay of
an inefficient hormonal regulation of the extrusion of a cell Na
load (Garay et al., 1980) and combine it with an inefficiency of

the sodium pump due to the modulation of the enzymatic complex by
cAMP that results in an increase of the pump activity to
compensate for the electrolyte load of the cells.

2. INFLUENCE OF CELL GROWTH ON ION TRANSPORT: RELATIONSHIP WITH THE ENERGETIC BALANCE

Over 50 years ago Warburg (1930) discovered that malignant
cancers ferment glucose to lactic acid much more rapidly than
most normal cells. In analyzing this phenomenon, Racker and
Spector (1981) correlated it to a high rate of ATP hydrolysis
that delivers the ADP and inorganic phosphate (Pi) required for
glycolysis. The source of ADP in most cells would then be an
inefficient Na/K ATPase that would result, to exert its normal
transporting functions, in an increased rate of ATP hydrolysis.
This surprising inversion of the general assumption by which a
correct pumping activity in the cells must be preceded by an
efficient phosphorylative oxidation that generates ATP was
strongly advanced but, after a few months, finally withdrawn by
the authors. In our opinion the main point of the above mentioned
theory (the involvement of Na/K ATPase during neoplastic growth)
is quite correct. What seems to be wrong is the explanation of a
primary role of the sodium pump in generating the substrates for
glycolysis.

What then is the implication of ATPase during cell growth?
To answer this question we have assayed the enzymatic activity in
several systems as follows:

1) Normal growth
   1.1 Chick embryo epidermis during embryonal development
   1.2 Human fibroblasts

2) Stimulated growth
   2.1 Chick embryo epidermis plus epidermal growth factor (EGF),
       a polypeptide hormone that enhances cell
       proliferation and differentiation (Savage et al., 1975;
       Frati et al., 1976)
   2.2 Human fibroblasts plus EGF
   2.3 Regenerating rat liver

3) Tumor growth
   3.1 HeLa cells in culture
   3.2 HeLa cells plus EGF
   3.3 Brain gliomas

Our results suggest a close correlation between the Na/K ATPase activity and cell growth (Verna and Frati, 1982a). It has been demonstrated that the enzymatic activity is always reduced when the growth rate is increased above normal. In addition this decrease is proportional to the malignancy of the growth rate.

On comparing results obtained, we found that normal cell growth shows only small variations of ATPase activity. The same systems stimulated with EGF show a reduced pumping activity (Verna and Frati, 1982, 1983a). This behavior depends on a double mechanism of action of the hormone, which first acts through the cAMP system (as most peptide hormones do), then reducing the -SH groups of the pump (Verna and Frati, 1982b, 1983a,b). Na/K ATPase activity decreases also during the first 4 hours of liver regeneration, showing a decreased sensitivity to exogenous addition of epinephrine and cAMP (Leoni et al., 1975). This behavior is probably due to a cAMP synthesis induced by the hormones involved during regeneration, which reduce the sensitivity of liver cells to the exogenous cAMP administration. So, also in this case, we have evidence that increased cAMP levels reduce the pumping activity.

The next step was to assay the enzymatic activity during tumor growth, because of the reduced cAMP levels in cancer cells. We always found a reduced Na/K ATPase activity (Verna and Frati, 1983b). Moreover, EGF addition, instead of decreasing the enzymatic activity in the first hour, increased it. This fact indicates that a raise in the cAMP level (as occurs during the action of most peptide hormones) restores the activity of the pump in cAMP-depleted cells. Furthermore, ATPase activity has been shown to play a fundamental role in the respiratory control of cerebral cortex (Lowry et al., 1954, Whittam and Blond, 1964). O'Connor (1969) and Laws and O'Connor (1970) suggested that in glioblastomas the vascular structure is altered, since the rich vascularization has not the influence on ATPase one would expect, given the rich oxygen supply. This finding supports Warburg's discovery and is in contrast with Racker's theory.

We therefore measured the Na/K ATPase activity as well as the cAMP and cGMP content in homogenates of human brain tumors. The measurements suggested a close correlation between the degree of malignancy and the enzymatic activity of the neoplastic cell (Verna et al., 1983c). On comparing the values obtained we found that cerebellar astrocytomas, notoriously benign, present the least depletion of Na/K ATPase if compared to normal cortex, glioblastomas the most, and the other tumor types depletion levels between these extremes. The degree of depletion is proportional to the degree of malignancy.

These results afford a striking example of the reduced biochemical activity that occurs during the metabolic changes characteristic of neoplastic cells. Moreover, we may regard the depletion of ATPase activity as a general observation applying to all tumor types, when we consider that oxidative phosphorylation is always markedly depressed in neoplastic cells.

In discussing our results we should remember that several papers have pointed out the role of cAMP and its related phosphorylations (Gottesman, 1980; Evain et al., 1979; Singh et al., 1981; Gottesman et al., 1981) in the control of cell growth. Also in this case the modulation of the Na/K ATPase by the protein kinase mediated phosphorylations postulated by Tria et al. (1974) could explain these results. According to this latter demonstration, high levels of cAMP should inhibit the activity of the sodium pump. The low levels of cAMP that have been found in many neoplastic cells would be expected to increase the activity of the sodium pump, and this fact would support Racker's theory. But we have always found this activity decreased in all the growing systems tested (Verna and Frati, 1982a,b; 1983a,b,d) and especially in the experiments on brain gliomas, with a close relationship to the malignancy and to the decrease of cAMP concentrations. How to explain this paradoxic effect?

Recently (Gottesman et al., 1980) several cell lines that are resistant to the growth inhibiting action of cAMP have been characterized. Their behavior is due to an alteration of the cAMP-dependent phosphorylations that are mediated by defective protein kinases. The subsequent demonstration that this alteration is due to a defective regulatory subunit (the one that binds cAMP) and results in a different cAMP request to activate the phosphorylative events in the cells leads us to postulate

that the dephosphorylative step of Na/K ATPase, necessary for the
activation of the enzyme, could be inefficient, thus resulting in
an inhibition of the ion transport system.

A further demonstration of the role of cAMP-dependent
phosphorylations in the modulation of the Na/K ATPase activity is
obtained by assaying the enzymatic activity in Chinese hamster
ovary (CHO) cells. Gottesman (1980) showed that treatment of
(CHO) cells with cAMP or its analogs resulted in changes in cell
shape (Hsie et al., 1971; Johnson et al., 1972; Van Veen et al.,
1976) and decreased nutrient transport (De Asua et al., 1974;
Gottesman, 1980; Hauschka et al., 1972; Kram et al., 1973; Paul,
1973; Rozengurt and Pardea, 1972) similar to those seen with
other fibroblastic cells.

To examine the mechanism of cAMP action, several independent
classes of mutant cell lines derived form the wild type CHO line,
which are resistant to the cell shape and growth inhibitory
effect of cAMP, have been isolated (Gottesman et al., 1980a).
These mutants provide a way to determine whether a specific cAMP
effect is mediated through the action of the cAMP-dependent
protein kinases or by a kinase-independent pathway.

In CHO 10001 (wild type) two peaks of PK activity are
present and Na/K ATPase is strongly reduced by the addition of
high cAMP concentrations (Verna et al., 1983e). CHO 10248 has a
defect in the protein kinases (only PKI is present) and does not
show modifications of Na/K ATPase activity at the same amounts of
cAMP at which it is affected in 10001. Furthermore a free RI (Rf)
is present in the mutant extract, not derived from RI and with a
different affinity for cAMP.

Thus, cAMP has different effects on 10248 as compared to
10001. We explain this behaviour as follows. The correct pumping
action of Na/K ATPase is known to be exerted by successive
coupling of phosphorylations and dephosphorylations of the
enzyme. According to this mechanism a normal protein kinase
activity is necessary for the phosphate transfer to the pump to
activate the phosphorylative step. But when cAMP content
increases, the consequent increase of cAMP-d-PK activity inhibits
the dephosphorylative step of the enzyme and results in a
decrease of the pumping activity. CHO 10001 cells are a striking
example of such behavior. In 10248, the presence of only PKI with

a defective RI and a higher affinity of RI not associated with holoenzyme (Rf) explains why the phosphorylative step of Na/K ATPase is inefficient at the same cAMP concentration at which it occurs in 10001. At higher concentrations, cAMP selectively binds to Rf (due to its higher affinity) so that the phosphotransferase activity of type I protein kinase cannot increase and does not inhibit the dephosphorylative step.

We were therefore able to show a direct correlation between cell proliferation and ion transport, expressed by the activity of the Na/K ATPase, and also that this correlation is exerted by the phosphorylative activity of the cAMP dependent protein kinases that are activated as a fourth step during the action of peptide hormones. Since this correlation is present in various kinds of cells, we postulate that the assay of the Na/k ATPase activity as well as the characterization of the protein kinase activities might be used as parameters of the response of the cell to peptide hormones and particularly of the transmission of the message inside the cell.

ACNOWLEDGEMENTS

This work was supported by C.N.R. progetto finalizzato "ONCOLOGIA". The authors also wish to thank Mrs. Wilma Savini and Mr. Sergio Ferraro for their helpful technical assistance. The CHO cell lines were kindly provided by Dr. M.M. Gottesman, N.I.H. Bethesda, Md. USA.

REFERENCES

Dahl, L.K., Heine M., and Tassinari, L. (1964). Effect of chronic excess of salt ingestion, vascular reactivity in two strains of rats with opposite genetic susceptibility to experimental hypertension. Circulation 30 (suppl.2), 11-22.

D'Armiento, M., Lacerna, F., Lauro, R., Modesti, A., Verna, R. and Ceccarelli, G., (1980). Effect of prazosin on the cAMP system in the spontaneously hypertensive rats (SHR) aorta. Eur. J. Pharm. 65, 243-247.

De Asua, L.J., Rozengurt, E.and Dulbecco, R. (1974). Kinetics of early changes in phosphate and uridine transport and cyclic AMP levels stimulated by serum in density-inhibited 3T3-cells. Proc. Natl. Acad. Sci. 71, 96-68.

Evain, D., Gottesman, M.M., Pastan, I., and Anderson, W. (1979). A mutation affecting the catalytic subunit of cAMP dependent protein kinase in CHO cells. J. Biol. Chem. 254, 6931-6937.

Frati, L., Cenci, G., Sbaraglia, G., Venza-Teti, D., and Covelli, I. (1976). Levels of epidermal growth factor in mouse tissues measured by a specific radioreceptor assay. Life Sci. 18, 905-912.

Garay, R.P. (1982). Inhibition of the Na/K cotransport by cAMP and intracellular Ca in human red cells. Bioch. Bioph. Acta 688, 6786

Garay, R.P., Dagher, G., Pernollet, M.G., Devynck, M.A., and Meyer, P., (1980). Inherited defect in Na/ K cotransport system in erythrocytes from essential hypertensive patients. Nature, Lond. 284, 281-283.

Garrahan, P., and Garay, R.P. (1976). The distinction between sequential and simultaneous models for sodium and potassium transport. Curr. Top. Membr. Transp. 8, 29-97.

Glynn, I.M. and Karlish, S. (1975). The sodium pump. Ann. Rev. Physiol. 37, 13-55.

Gottesman, M.M., Le Cam, A., Bukowsky, M., and Pastan, I. (1980). Isolation of multiple mutants of CHO. Somatic Cell Genetics. 6, 45-61.

Gottesman, M.M. (1980). Genetic approaches to cAMP effects in cultured mammalian cells. Cell 22, 329-330.

Gottesman, M.M., Singh, T., Le Cam, A., Roth, C., Nicolas, J.C., Cabral, F. and Pastan, I. (1981). cAMP dependent phosphorylation in cultured fibroblasts: A genetic approach. Cold Spring Harbor Conferences on Cell Proliferation, Vol. 8, Protein Phosphorylation, pp. 195-209.

Hauschka, P.V., Everhart, L.T., and Rubin, R.W. (1972). Alteration of nucleoside trasport of Chinese hamster cell by dibutyryl adenosine 3':5'cyclic monosphosphate (thymidine and uridine uptake/thymidine kinase/DNA and RNA synthesis). Proc. Natl. Acad. Sci. USA 69, 3542-3546

Hsie, A.W., Jones, C. and Puck, T.T. (1971). Further changes in differentiation state accompanying the conversion of Chinese hamster cells of fibroblastic form by adenosine cyclic 3':5' monophosphate and hormones. Proc. Natl. Acad. Sci. USA 68, 1648-1652.

Hsie A.W., Jones C. and Puck T.T. (1971). Morphological transformation of Chinese Hamster Cells by dibutyryl adenosine cyclic 3':5' monosphophate and testosterone. Proc. Natl. Acad. Sci. USA 68, 358-361.

Johnson, G.S., Friedman, R.M. and Pastan, I. (1971). Restoration of several morphological characteristics of normal fibroblasts in sarcoma cells treated with adenosine 3':5' cyclic monophospate and its derivatives. Proc. Natl. Acad. Sci. USA 68, 425-429.

Kram, R., Mamont, P. and Tomkins, G.M. (1973). Pleiotypic control by adenosine 3':5' Cyclic monophosphate: A model for growth control in animal cell. Proc. Natl. Acad. Sci. USA 70, 1432-1436.

Laws, E.R. and O'Connor, J. (1970). ATPase in human brain tumors. J. Neurosurgery 33, 167-171.

Leoni, S.,Luly, P., Mangiantini, M.T., Spagnuolo, S., Trentalance, A., and Verna, R. (1975). Hormone responsiveness of plasma membrane bound enzymes in normal and regenerating rat liver. Biochim. Biophys. Acta 394, 317-322.

Lowry, O.H., Roberts, N.R., Wu, M.L., Hixon, W.S., Crawford, E.J. (1954). Quantitative histochemistry of brain. II: Enzyme measurement. J. Biol. Chem. 207, 19-37.

O'Connor, J.S. and Laws, E.R. (1963). Histochemical survey of brain tumor enzymes. Arch. Neurol. 9, 641-651.

O'Connor, J.S. and Laws, E.R. (1969). Changes in histochemical staining of brain Tumors blood vessels associated with increasing malignancy. Acta Neuropathol. 14, 616-673.

Paul, D. (1973). Quiescent SV40 virus transformed 3T3 cells in culture. Biochem. Biophys. Res. Commun. 53, 745-753

Racker, E. and Spector M. (1981). Warburg effect revisited: Merger of Biochemistry and Molecular Biology. Science, 213, 303-307.

Rozengurt, E., and Pardea, A.B. (1972). Opposite effect of dibutyryl cAMP and serum on growth of CHO. J. Cell. Physiol. 80, 273-280.

Ryan, W.L. and Heidrick, M.L. (1968). Inhibition of cell growth in vitro by cAMP. Science 162, 1484-1485.

Savage, G.R., Inagami, T. and Cohen S. (1975). The primary structure of epidermal growth factor. J. Biol. Chem. 247, 7612-7621.

Singh, T.J., Roth, C., Gottesman, M.M. and Pastan, I. (1981). Characterization of cAMP resistant CHO cell mutants lacking type I protein kinase. J. Biol. Chem. 256, 926-932.

Verna, R. and Frati, L. (1982a). Na/K ATPase and cell growth: A
    marker of neoplastic growth? In: Membranes in Tumor Growth
    T. Galeotti et al., Eds. Elsevier North Holland pp.515-525.

Verna, R. and Frati, L. (1982b). Na/K ATPase and cell growth: EGF
    modulates enzymatic activity in cultured fibroblasts.
    Cytobios 35 181-186.

Verna, R. and Frati, L. (1983a). Na/K ATPase and cell growth:
    Effect of epidermal growth factor on the enzymatic activity
    in chick embryo epidermis during the embryonal development.
    Int. J. Biochem. 15, (1), 1-3.

Verna, R. and Frati, L. (1983b). Na/K ATPase and cell growth.
    III: Enzymatic activity in cultured and EGF stimulated HeLa
    cells. Int. J. Biochem., 15 (2), 137-138.

Verna, R. and Muraro, R. (1983c). Essential hypertension: Role of
    Na/K ATP-A and relationship with the cyclic nucleotides
    system. Cell. Mol. Biol. 29 (1), 93-102.

Verna, R., Nardi, P., Muraro, R., Delfini, R., Frati, L. (1983d).
    Na/K ATPase and cell growth IV: A metabolic marker of human
    brain tumors? J. Neuro. Sci. 27 (2) 77-82.

Warburg, O. (1930). The metabolism of tumors. Constable & Co.,
    London, 327.

Whittam, R. and Blond, D.M. (1964). Respiratory control by an
    ATPase involved in active transport in brain cortex.
    Biochem. J. 92, 147-158.

REGULATION OF Na$^+$ AND K$^+$ TRANSPORT IN MACROPHAGES

Pierre Braquet[1], Javier Diez[2], Francis V. DeFeudis[1],
Roberto Verna[3], and Ricardo Garay[2]

[1]IBH Laboratories, 17 Avenue Descartes, 92350
Le Plessis Robinson, France
[2]U7 INSERM/LA 318 CNRS, Hôpital Necker, Paris, France
[3]Istituto di Patologia Generale, Via le Regina Elena
324, 00161 Roma, Italy

INTRODUCTION

Knowledge of Na$^+$ and K$^+$ metabolism at a molecular level is not so advanced as that of oxygen, sugar or lipid metabolism. This might be related to the fact that all known functional proteins of Na$^+$ and K$^+$ metabolism are membrane transport proteins. One successful approach toward elucidating the molecular mechanism of Na$^+$ and K$^+$ transport is to study transport kinetics. This approach provides information concerning the dependence of cation flux on the chemical composition on either side of the membrane and on other variables (e.g., membrane potential, temperature).

Recently, the kinetic characterization of different Na$^+$ and K$^+$ transport systems in human red blood cells[1-3] has been applied to studies of the physiology and physiopathology of Na$^+$ and K$^+$ metabolism. Na$^+$ and K$^+$ transport systems are less well characterized in other cells, including macrophages. However, some recent observations have indicated that transmembrane ion movements might be primary events for secretory and phagocytic functions of macrophages and polymorphonuclear leukocytes. Membrane depolarization seems to precede the release of superoxide anion induced by zymosan and other chemical agents in rat alveolar macrophages[4] and neutrophils[5].

In addition, the stimulation of human macrophages from peripheral blood[6] and of rabbit polymorphonuclear leukocytes[7] by chemotactic factors induces changes in membrane potential and in transmembrane cation fluxes.

The resting membrane potential of macrophages depends mainly on the transmembrane K^+ permeability, which is heavily dependent on the cytoplasmic free Ca^{2+} concentration[8]. An increased Na^+ influx seems to initiate membrane depolarization[4]. The transmembrane Na^+ gradient appears to provide the energy for net entry of glucose and amino acids into macrophages[4]. Although Na^+ and K^+ movements appear to play an important role in many vital functions of macrophages, little information is available concerning Na^+ and K^+ transport systems and and their regulation in these cells. Indeed, only a ouabain-sensitive Na^+, K^+-pump[9] and Ca^{2+}-dependent, K^+-permeability (Gardos effect)[10,11] have been examined. Thus, we decided to investigate different Na^+ and K^+ transport systems in thioglycollate-elicited mouse peritoneal macrophages and their possible regulation by icosanoids.

MATERIALS AND METHODS

Animals

Female mice, 5 - 8 weeks old, weighing 20 - 30 g, of the 57 BL/S [(H-2^b)] and DBA/2 [(H-2^d)] inbred strains (Centre de Sélection et d'Elevage des Animaux de Laboratoire, Orléans, France), were used. They had free access to water and a standard diet during the experimental period.

Preparation of cells

Peritoneal macrophages were elicited by injection into the peritoneal cavity of 2.5 - 3 ml of sterile thioglycollate medium (Institut Pasteur, Paris, France). The cells were collected 3 - 5 days later by washing the peritoneal cavity with Hank's balanced salt solution (HBSS). About 10^7 cells were collected per mouse, and more than 80% of these presented the morphologic aspect of macrophages ; 5 - 10 mice were used for each experiment. Cell suspensions were centrifuged at 4°C (3750 g for 4 min), and the supernatants were discarded. Macrophages were then washed once in $MgCl_2$ solution (110 mM).

Measurement of cation fluxes and intracellular ion concentrations

Na^+ and K^+ fluxes were determined in fresh macrophages according to previously published methods for human erythrocytes[3,12,13]. Briefly, Na^+ and K^+-pump fluxes are represented by the fraction of total Na^+ efflux in a Mg^{2+}-sucrose medium ($MgCl_2$ 75 mM, sucrose 85 mM, glucose 10 mM, MOPS-Tris 10 mM, pH 7.4 - 7.5, osmolarity

295 - 300 mOs-moles/l) that is inhibited by ouabain (1 mM); Na^+, K^+-cotransport fluxes are represented by the fraction of Na^+ and K^+ efflux in Mg^{2+}-sucrose-ouabain medium that is inhibited by 0.02 mM bumetanide; passive Na^+ and K^+ permeabilities are represented by the ouabain- and bumetanide-resistant fraction of Na and K^+ efflux in a Mg^{2+}-sucrose medium. The ouabain- and bumetanide-resistant fraction of Na^+ efflux in a Mg^{2+}-sucrose medium in the presence of LiCl (10 mM) was considered as a Na^+-Li^+ exchange mechanism. The ouabain- and bumetanide resistant fraction of Na^+ efflux in Mg^{2+}-sucrose medium in the presence of $CaCl_2$ (1 mM) was considered to represent a Na^+-Ca^{2+} exchange mechanism. The existence of a Na^+-H^+ or K^+-H^+ exchange mechanism was studied by incubating macrophages in an unbuffered medium (NaCl 150 mM or KCl 150 mM, $MgCl_2$ 1 mM, glucose 5 mM, ouabain 1 mM, bumetanide 0.02 mM, DIDS 0.01 mM) and by then measuring external pH.

To measure intracellular Na^+ and K^+, cells were suspended in a Mg^{2+}-sucrose solution ; 0.1 ml of the cell suspension was added to 4 ml of acationox (0.2 %) solution and then frozen. Cells were lysed by rapid thawing. Na^+ and K^+ were measured with an Eppendorf flame photometer. The effects of arachidonic acid (AA) and of various icosanoids, including PGE_1 , PGE_2 , PGI_2 , 6-keto-$PGF_{1\alpha}$, $PGF_{2\alpha}$ and leukotrienes (LT) B_4 , C_4 , D_4 on cation fluxes and intracellular ion concentrations were also determined.

RESULTS

Na⁺ and K⁺ contents of macrophages

The method used to determine the Na^+ and K^+ contents of thioglycollate-elicited peritoneal macrophages has two main sources of artifact: (i) the Na^+ , K^+-pump and other active mechanisms are partially inhibited in the cold; thus, Na^+ tends to enter into the cells during the procedure of drawing and pooling the peritoneal exudates in cold Hank's solutions, and (ii) the washing of cells in Na^+ and K^+ free media may enhance the release of both cations from the cells. To study the influence of such factors, some control experiments were conducted. We first verified that the starving of peritoneal exudates in cold Hank's solution for no more than 30 minutes did not alter significantly the measured Na^+ and K^+ contents. Secondly, it was found that the presence of ouabain and ions of the washing solution did not alter significantly the measurement of Na^+ and K^+ contents (Table 1). Conversely, cell Na^+ content tended to decrease slightly as the number of washings was increased (Table 1).

Table 1. Intracellular Na^+ and K^+ Concentrations in Mouse Macrophages.

Washing medium	Number of washings	Na^+_i (mmol/l cells)	K^+_i (mmol/l cells)
$MgCl_2$, 110 mM	1	19.6 ± 2.6 (7)	70.2 ± 9.5 (7)
$MgCl_2$, 110 mM	2	14.1 ± 2.7 (3)	62.0 ± 10.0 (3)
$MgCl_2$, 110 mM	3	11.0	64.0
$MgCl_2$, 110 mM Ouabain, 0.1	1	17.1 ± 1.3 (2)	60.7 ± 6.1 (2)
Choline-Cl , 150 mM	1	24.0 ± 3.8 (3)	51.7 ± 19.7 (3)
$Mg(NO_3)_2$, 110 mM	1	14.2	50.7

Values are given as means ± SD or as individual values; numbers of experiments are indicated in parentheses.

Table 2. Effects of K^+, Ouabain, Bumetanide and Ca^{2+} on Na^+ and K^+ Fluxes in Mouse Macrophages

Transport system	Flux component	Na^+ Efflux mmol/l cells x h	K^+ Efflux mmol/l cells x h
Na^+, K-pump	K^+-stimulated Ouabain-sensitive	7.2 ± 2.4 (12)	
Na^+, K^+-cotransport	Bumetanide-sensitive	5.1 ± 2.3 (10)	11.2 ± 3.6 (12)
Na^+, Ca^{2+}-exchange	Ca^{2+}-stimulated	5.6 ± 3.7 (6)	
Uncharacterized	Ouabain- and bumetanide resistant	20.1 ± 8.2 (12)	74.9 ± 56.2 (12)

Fluxes were studied in Mg^{2+}-sucrose medium. Values are given as means ± SD; numbers of experiments are indicated in parentheses.

Na$^+$ and K$^+$ effluxes in Mg^{2+}-sucrose medium. The effects of ouabain, bumetanide, K$^+$, Li$^+$ and Ca^{2+}

Na$^+$ and K$^+$ effluxes might be catalyzed by different transport systems. The use of selective inhibitors or transport-stimulating ions permitted analysis of the fractional flux catalyzed by each system. Ouabain selectively blocks the exchange of internal Na$^+$ for external K$^+$ that is catalyzed by Na$^+$, K$^+$-pump[1]. Bumetanide, at low doses, is a quasi-selective inhibitor of fluxes catalyzed by the Na$^+$, K$^+$-cotransport system[3]. Li$^+$ and Ca^{2+} may activate Na$^+$-Li$^+$ and Na$^+$-Ca^{2+} exchange mechanisms, respectively[2,14].

Figure 1 shows Na$^+$ and K$^+$ effluxes from macrophages into three different media: (i) K$^+$ medium; (ii) ouabain medium; (iii) ouabain + bumetanide medium. The difference between the first two media is a measure of the Na$^+$ efflux catalyzed by the Na$^+$, K$^+$-pump. The difference between media 2 and 3 is a measure of outward Na$^+$, K$^+$-cotransport. All of these cation fluxes deviated from monoexponential kinetics after 5 - 30 minutes of incubation (Fig 1). In 12 different experiments the ouabain-sensitive Na$^+$ efflux was 4 - 10 m-moles/l cells x h (Table 2). The bumetanide-sensitive Na$^+$ efflux was 2.5 - 7.6 m-moles/l cells x h, and the bumetanide-sensitive K$^+$ efflux was 7 - 15 m-moles/l cells x h (Table 2). However, most of the total Na$^+$ and K$^+$ effluxes were resistant to ouabain and bumetanide (Table 2). In addition, the ouabain- and bumetanide-resistant K$^+$ efflux was extremely variable from one experiment to the other. Li$^+$ showed no effect on the ouabain- and bumetanide-resistant Na$^+$ efflux from macrophages (Table 3).

Table 3.   Effect of Li$^+$ on Ouabain- and Bumetanide-Resistant Na$^+$ Efflux

Experimental Conditions	Na$^+$ Efflux mmol/l cells x h
Control	28·3
Li$^+$, 5 mM	28·0
Li$^+$, 10 mM	25·2

Experiments were performed in Mg^{2+}-sucrose medium. Values are given as the means of two experiments.

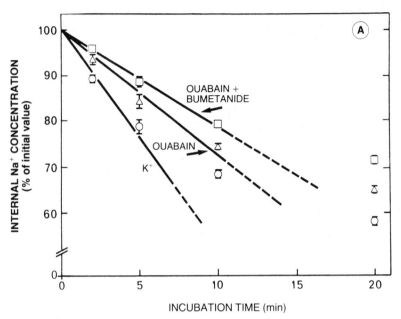

Fig. 1a. Na$^+$ efflux from thioglycollate-elicited mouse macrophages.
          Ouabain- and bumetanide + ouabain-sensitive components
          are indicated.

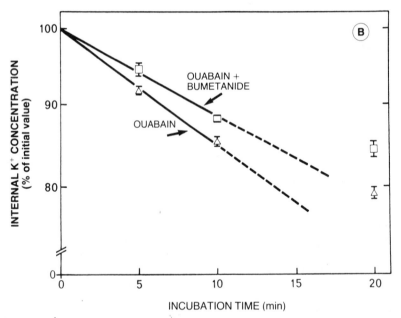

Fig. 1b. K$^+$ efflux from thioglycollate-elicited mouse macrophages.
          Ouabain- and bumetanide + ouabain-sensitive components
          are indicated.

External Ca^{2+} was able to stimulate ouabain- and bumetanide-resistant Na^+ efflux. The magnitude of this Ca^{2+}-stimulated Na^+ efflux varied from 2 to 9 m-moles/l cells x h (Table 2). In these experiments the reference Mg^{2+}-sucrose medium contained between 2 and 8 μM Ca^{2+}. Additional experiments, in which Ca^{2+} was buffered with EGTA, revealed that activation of Na^+ efflux occurs between 10 and 100 μM free Ca^{2+} (data not shown).

Ouabain- and bumetanide-resistant K^+ fluxes. The effects of N-ethylmaleimide (NEM) and osmolarity

The membrane potential of macrophages depends mainly upon K^+ permeability. In two experiments, in which K^+ effluxes was measured in Mg^{2+}-sucrose medium, increases in osmolarity above physiological levels strongly inhibited K^+ efflux (data not shown). Also, NEM inhibited K^+ efflux in isotonic, but not in hypertonic, medium, indicating that hypertonicity and NEM could inhibit the same K^+ transport system (Fig. 2). NEM (2 mM) enhanced K^+ efflux in both isotonic and hypertonic media, indicating a toxic phenomenon (Fig. 2).

Transmembrane H^+ exchange in macrophages. Effects of external Na^+ and K^+

When macrophages were incubated in unbuffered media (isotonic NaCl or isotonic KCl), they tended to buffer the external pH between pH 6.5 and 7.5, this effect being similar in Na^+ and K^+ media (data not shown).

The effects of AA and of icosanoids on Na^+ and K^+ transport systems in the macrophage

Arachidonic acid (10^{-5} M) stimulated Na^+ efflux from macrophages, this effect being maximal at 15 min. Conversely, indomethacin (10^{-5} M) inhibited Na^+ efflux from these cells (Fig. 3). A similar inhibitory effect was observed in the presence of AA + indomethacin, indicating that AA and its metabolites might be involved in the above-mentioned stimulation of Na^+ efflux (Fig. 3).

Table 4 shows the effects of PGE_1, PGE_2, PGI_2, $PGF_{2\alpha}$, and 6-keto-$PGF_{1\alpha}$ on several Na^+ and K^+ transport systems of the macrophage. At a very high concentration (10^{-5} M), all prostaglandins that were tested inhibited outward fluxes catalyzed by the Na^+, K^+-cotransport system. A significant inhibition of both Na^+ and K^+ effluxes was observed in the presence of PGE_1. In addition, PGE_2 and PGI_2 significantly inhibited outward K^+ and outward Na^+ fluxes, respectively. PGE_1 was the only prostaglandin tested that exerted a significant effect on the Na^+-Ca^{2+} exchange process; i.e., it increased the stimulation of Na^+ efflux by external Ca^{2+} into a Na^+-free medium (Table 4). At 10^{-5} M, PGE_1, PGE_2 and PGI_2

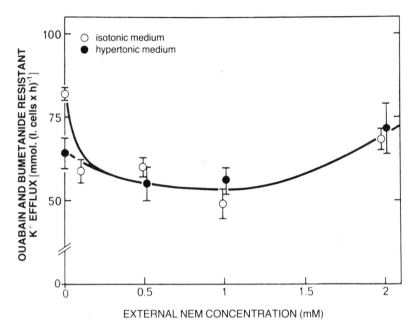

Fig. 2.   Dose-dependent inhibition of ouabain- and bumetanide-
          resistant K$^+$ efflux by N-ethylmaleimide (NEM) in the
          macrophage.

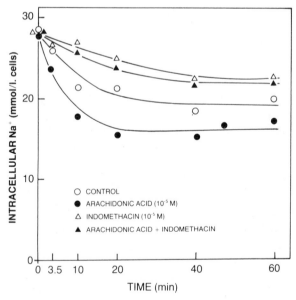

Fig. 3.   Effects of exogenous arachidonic acid and of indomethacin
          on Na$^+$ efflux in the macrophage.

tended to increase the Na^+ efflux catalyzed by the Na^+, K^+-pump (Table 4).

As shown in Figure 4, prostaglandins (at 10^{-5} M) did not significantly affect ouabain- and bumetanide-resistant Na^+, K^+ effluxes. Regarding the effects of lipoxygenase products on macrophage Na^+ and K^+ transport, it was found that LTB_4 (10^{-5} M) increased both total Na^+ efflux and total K^+ efflux, whereas LTC_4 and LTD_4 were ineffective (Fig. 5).

DISCUSSION

The results shown herein indicate that the Na^+, K^+-pump of macrophages catalyzes a small fraction of the total cation movements across the cell membrane. This effect is opposite to that which occurs in the human red blood cell membrane in which the Na^+, K^+-pump catalyzes almost 80% of the total Na^+ efflux or K^+ influx[1]. The stimulation of Na^+ efflux by external Ca^{2+} (Table 2) suggests the presence of a Na^+-Ca^{2+}-exchange mechanism in the macrophages studied. Nevertheless, a Na^+-Ca^{2+}-exchange mechanism might cooperate with the Ca^{2+}-pump in order to extrude the Na^+ load during macrophage activation[4]. Interestingly, we have not detected Na^+, Li^+-counter-transport in macrophages, further indicating that in erythrocytes this system may be a vestigium of Na^+-Ca^{2+}-exchange[15].

No evidence for Na^+-H^+-exchange could be observed in the studies reported here, a finding which was unexpected since acidification occurs during phagocytosis[16]. Thus, it is possible that Na^+-H^+-exchange might occur only after macrophage activation, or that our methods were not sensitive enough to detect it. In contrast to findings obtained with erythrocytes[17], we have found that NEM inhibits K^+ fluxes in macrophages incubated in isotonic medium. A similar inhibition has been found in leukocytes (P.L. Lauf, unpublished observations, 1983). Taken together, these observations suggest that the regulation of cell volume might be mediated by different basic mechanisms in different types of cells.

One of the most important effects described herein was that K^+ efflux is strongly inhibited by an hypertonic medium. This indicates that increases in osmolarity might lead to membrane depolarization and macrophage activation. Considering the hypertonicity of inflammatory foci[18], this effect might be one of the physiological signals involved in macrophage activation.

It has been suggested that the turnover of transmembrane cation translocation is under hormonal control[19,20]. Certain hormones (e.g., catecholamines, glucagon, insulin and thyroxine) might modify the activity of the Na^+, K^+-pump[21,23] and the Na^+, K^+ cotransport system[24]. Prostaglandins, like some other hormones and neurotransmitters, can stimulate or inhibit adenylate cyclase, depending on the cell and the prostaglandin receptor type studied.

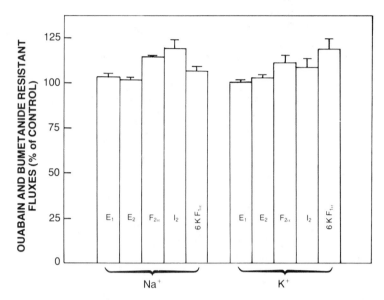

Fig. 4. Effects of various prostaglandins on ouabain- and
bumetanide-resistant Na^+ and K^+ fluxes in the macrophage.

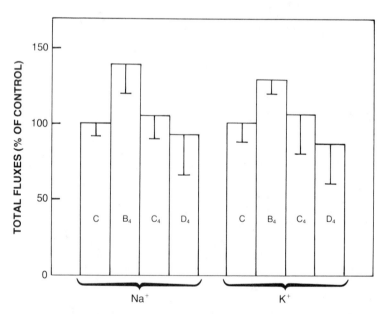

Fig. 5. Effects of various leukotrienes on total Na^+ and K^+
fluxes in the macrophage.

Table 4. Effects of Prostaglandins on Na^+ and K^+ Transport Systems of Mouse Macrophages

Transport System	Control	PGE_1	PGE_2	PGI_2	$PGF_{2\alpha}$	6-keto-$PGF_{1\alpha}$
Na^+,K^+ co-transport						
Na^+ efflux	5.1 ± 2.3	0.3 ± 0.2	2.4 ± 0.9	2.1 ± 0.2	3.3 ± 1.9	2.1 ± 1.6
K^+ efflux	14.2 ± 3.6	1.2 ± 0.1	0.8 ± 0.6	5.5 ± 4.0	4.4 ± 3.5	4.4 ± 4.0
Na^+,Ca^{2+} exchange	3.7 ± 2.1	7.4 ± 2.4	2.5 ± 0.6	5.3 ± 0.9	5.0 ± 1.4	4.4 ± 1.1
Na^+,K^+-pump	6.5 ± 4.6	8.6 ± 5.9	7.6 ± 4	9.8 ± 6.6	7.1 ± 5.3	6.6 ± 5.0
		(+30%)	(+17%)	(+50%)		

Values are given as means ± SD of several experiments. Values are expressed as m-mole $(l.cells \times h)^{-1}$ Prostaglandins were assayed at 10^{-5} M concentration.

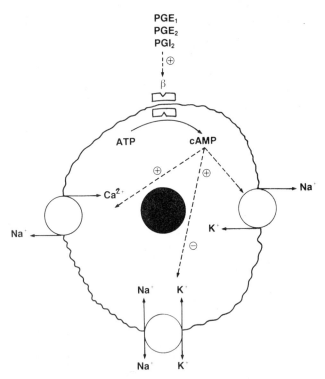

Fig. 6.   A model of the regulation of ion fluxes by prostaglandins via activation of adenylate cyclase.

Generally, in elicited populations of peritoneal macrophages, prostaglandins, mainly PGE_1 , PGE_2 , and PGI_2 increase intracellular cAMP levels[25,26]. As cAMP has been considered as an intermediate for the hormonal control of Na+ and K+ transport[24,27,28], we propose that in macrophages certain prostaglandins (PGE_1 , PGE_2 and PGI_2) might modulate Na+ and K+ transport systems by stimulating adenylate cyclase (Fig. 6). The net effect might be related to an increase in the electrochemical Na+ gradient.

It has been shown that cAMP[29] and prostaglandins, mainly PGE_2[26], regulate some macrophage functions. Also, considerable evidence indicates a relationship among transmembrane potential changes, transmembrane ionic movements and the control of the phagocytic response of the macrophage[4]. Thus, it might be interesting to consider whether prostaglandins might regulate macrophage functions by modulating the electrochemical Na+ gradient. Also, the stimulatory effect of PGI_2 on the macrophage Na+, K+-pump, if observable in smooth mucle cells, might contribute to our knowledge of the regulation of vascular tone since PGI_2 has vasodilator properties[30]. Further investigations are also required to explain whether PGE_1 ,which stimulates Na+-Ca2+-exchange in macrophages, might contribute to the regulation of vascular tone through its effects on the vascular Na+-Ca2+-exchange mechanism that has been postulated by Blaustein[31].

The finding that LTB_4, which has been proposed as a Ca2+-ionophore[32], enhanced both Na+ and K+ effluxes from macrophages (see Fig. 5) indicates that this leukotriene could be involved in regulating Na+ and K+ transport.

In conclusion, macrophages contain most of the cation transport systems previously described for other cells, but the properties of these systems differ from those of other cells, particularly erythrocytes. Further study of such systems under different experimental conditions might clarify the basic mechanisms involved in initiating and modulating macrophage activation and function.

REFERENCES

1. R.P. Garay and P. Garrahan, The interaction of sodium and potassium with the sodium pump in red cells, J. Physiol. (Lond.), 231: 297-325, (1973).

2. B. Sarkadi, J.K. Alifimoff, R.H. Gunn and D.C. Tosteson, Kinetics and stoichiometry of Na+-dependent-Li+ transport in human red blood cells, J. Gen. Physiol; 72: 249-265, (1978).

3. R.P. Garay, C. Nazaret, P. Hannaert and M. Price,
    Abnormal Na^+, K^+-cotransport function in a group of pa-
    tients with essential hypertension, Eur.J.Clin.Invest;
    13: 311-320, (1982).

4. P.R. Miles, L. Bowman and V. Castranova, Transmembrane
    potential changes during phagocytosis in rat alveolar
    macrophages, J.Cell.Physiol; 106: 109-117, (1981).

5. G.S. Jones, K.Van Dyke and V. Castranova, Transmembrane
    potential changes associated with superoxide release
    from human granulocytes, J.Cell.Physiol; 106: 75-83,
    (1981).

6. E.K. Gallin and J.I. Gallin, Interaction of chemotactic
    factors with human macrophages, J.Cell.Biol; 75:
    277-289, (1977).

7. P.E.R. Tatham, P.J. Delves, L.Shen and I.M. Roitt,
    Chemotactic factor-induced membrane potential changes
    in rabbit neutophils monitored by the fluorescent dye
    3,3'-dipropylthiadicarbocyanine iodide, Biochim.Bio-
    phys.Acta, 602: 285-298, (1980).

8. G.M. Oliveira-Castro and G.A. Dos Reis, Electrophysio-
    logy of phagocytic membranes. III. Evidence for a
    calcium-dependent potassium permeability changing du-
    ring slow hyperpolarization of activated macrophages,
    Biochim.Biophys. Acta, 640(2): 500-511, 1981).

9. C.E. Cross, M.G. Mustafa, P. Peterson and J.A. Hardie,
    Pulmonary alveolar macrophage. Membrane sodium ion,
    potassium ion, and magnesium ion adenosine triphospha-
    tase system, Arch.Intern.Med., 127: 1069-1077, (1971).

10. A. Holian and R.P. Daniele, The role of calcium in the
    initiation of superoxide release from alveolar macro-
    phages, J.Cell.Physiol., 113: 87-93, (1982).

11. A. Holian and R.P. Daniele, Formyl-peptide stimulation
    of superoxide anion release from lung macrophages:
    sodium and potassium involvement, J.Cell.Physiol.,
    113: 413-419, (1982).

12. D.Cusi and R.P.Garay, The effect of tienilic acid on
    Na^+ and K^+ transport in human red cells, Mol. Pharmacol.
    19: 438-443, (1981).

13. R.P. Garay, G. Dagher, M.G. Pernollet, M.A. Devynck and P. Meyer, Inherited defect in a Na^+, K^+-cotransport system in erythrocytes from essential hypertensive patients, Nature (Lond), 284: 281-283, (1980).

14. M.P. Blaustein, Sodium ions, calcium ions, blood pressure regulation and hypertension: a reassessment and a hypothesis, Am.J.Physiol., 232: C165-75, (1977).

15. D.C. Tosteson, Ion transport across biological membranes. cation counter-transport and cotransport in human red cell membranes. in: "Cell Membrane in Function and Dysfunction of Vascular Tissue", T. Godfraind and P. Meyer, eds., pp. 145-158 (1981). Elsevier/North-Holland (Amsterdam).

16. M.S. Jensen and D.F. Bainton, Temporal changes in pH with the phagocytic vacuole of the polymorphonuclear neutrophylic leukocyte, J.Cell.Physiol., 56:379-388, (1973).

17. P.K. Lauf and B.E. Theg, A chloride-dependent K^+ flux induced by n-ethylmaleimide in genetically low K^+ sheep and goat erythrocytes, Biochem.Biophys,Res.Comm., 92: 1422-1428, (1980).

18. H.Z. Movat, The acute inflammatory reaction, in: Inflammation, Immunity and Hypersensitivity, H.Z. Movat ed., pp. 1 - 162. ,Harper and Row. Pub. ,1979.

19. J.J. Granthamm, The renal Na^+ pump and vanadate, Am.J. Physiol. ,239: F97 - F106 (1980).

20 K.J. Sweadner and J.M. Goldin, Active transport of sodium and potassium ions : mechanism, function and regulation, New Eng. J. Med. ,302: 777-83 (1980).

21 R. Biron, A. Burger, A. Chinet, T. Clausen, and F. Dubois, Thyroid hormones and the energetics of active sodium-potassium transport in mammalian skeletal muscles, J Physiol. (Lond.) 297: 47-60 (1979).

22. T. Clausen and J. Flatman , The effect of catecholamines on Na^+, K^+-transport and membrane potential in rat soleus muscle, J Physiol. (Lond.) 270: 383-414 (1977).

23. T. Clausen and O. Hansen , Active Na^+, K^+-transport and the role of ouabain binding. The effect of insulin and other stimuli on skeletal muscle and adipocytes, J Physiol. (Lond.) 270: 415-430 (1977).

24. R.P. Garay, Inhibition of the Na^+ K^+ cotransport system
     by cyclic AMP and intracellular Ca^{2+} in human red cell,
     Biochim. Biophys. Acta., 688:786-92 (1982).

25. J.L. Bonta and M.J.P. Adolfs, Interaction between pros-
     taglandin E_2 and prostacyclin in regulating levels of
     cyclic AMP in elicited populations of peritoneal macro-
     phages, in: "Advances in Prostaglandin, Thromboxane and
     Leukotriene research" vol.12, B. Samuelson, R. Paoletti
     and P Ramwell, eds., pp. 13-17, Raven Press, New York
     (1983).

26. B. Gemsa, Stimulation of PGE released from macrophages
     and possible role in the immune response. in: "Lympho-
     kine", vol.4, E. Pick, ed., pp. 335-75. Academic Press,
     New York (1981).

27. J.L. Alper, M.C. Palfrey, S.A. De Riemer and
     P. Greengard, Hormonal control of protein phosphoryla-
     tion in turkey erytrhocytes, J Biol.Chem. 255: 11029-
     11039 (1980).

28. P. Luly, O. Barnabei and E. Tria, Biochim. Biophys.
     Acta, 282: 447-52 (1972).

29. J.G. Zendegui and T.W. Klein, Reduction in cyclic
     3', 5' - adenosine monophosphate diesterase activity
     in exudate and cultured mouse peritoneal macrophage,
     J. Reticuloendothelial Society , 31: 455-67 (1982).

30. E.J. Spokas, J. Quilley and J.C. Mc Giff, Hypertension:
     Physiopathology and treatment, in : "Hypertension",
     G. Genest, O. Kuchel, P. Hamet and Cantin, eds.,
     pp. 373-383. Mc Graw Hill, New York. (1983).

31. M.P. Blaustein, Sodium ions, calcium ions, blood pres-
     sure regulation and hypertension: a reassessment and
     hypothesis, Am.J.Physiol., 232: 165-73 (1977).

32. C.N. Serhan, V.J. Fridovich, E.J. Goetzl, B.D. Dunhan
     and G. Weissmann, Leukotriene B_4 and phosphatidic acid
     are calcium ionophores, J. Biol. Chem., 1257: 47 46-52
     (1982).

CHRONOBIOLOGY OF Na/K ATPase ACTIVITY IN HUMAN ERYTHROCYTE MEMBRANES

*Roberto Verna,**Pierre Braquet,*Pietro Cugini
and ***Ricardo Garay

Institute of General Pathology, University of Rome, viale
Regina Elena, 324 - 00161 Roma, Italy (*)

I.H.B. Laboratories, Le Plessis Robinson, France (**)

INSERM U7, Hôpital Necker, Paris, France (***)

INTRODUCTION

Wide interest in biological rhythms has developed during
recent years, given the great physiological importance of this
phenomenon, and it is now generally accepted that chronobiology
has a number of pathogenic and therapeutic implications. Since
most of these implications are due to the interaction of
physiological and pharmacological substances with the plasma
membrane, and the plasma membrane itself is the fundamental
structure which regulates the interrelationships between cells,
we have investigated the possibility of a dependence of this
structure on the biological rhythms or, in other words, the
possibility of a fluctuation of the membrane structure.

As a first approach to this investigation, we have studied
the behavior of Na/K ATPase activity during a 24 h span in human
erythrocyte membranes. The choice of red blood cells was
determined by the particular structure of the erythrocytes (lack
of nucleus and of cytoplasmic organelles, lack of other energy
dispersing systems except the sodium pump), because of their
availability and because of the fundamental role of the sodium
pump in maintaining their size and shape.

MATERIAL AND METHODS

Investigation was performed in ten clinically healthy
subjects. Informed consent was obtained from each subject
investigated. The research was planned according to a
chronobiologic design.

Subjects were sampled by drawing venous blood on a
heparinized syringe at 6:00, 8:00, 12:00, 18:00, 20:00 and 24:00
in recumbent position after a week-long synchronization to a
light-dark regimen (light on: 7:00, light off: 23:00) and meal
schedule (breakfast 8:00; lunch 12:00; dinner 18:00). The diet
was habitual in sodium (1.7-2 mEq/Kg/24 h) and potassium
(0.7-1mEq/Kg/24 h) and dietary intake of carbohydrates, fats and
proteins per kg/24 h was of 3 g, 1 g and 1 g respectively. The
investigation was started when urinary excretion of sodium and
potassium in subjects under investigation was found to be
proportional to the amount ingested over the 24 h span.

Erythrocyte membrane purification was carried out following
the method described by Hanahan (Hanahan and Ekholm, 1974)
modified as follows: blood was centrifuged at 1000 rpm and plasma
discarded; two more washings with isotonic Tris buffer (310 mOsm
pH 7.6) were performed at 1000 rpm in a J-21 Beckman centrifuge
at 4°C. Hypotonic buffer was between 7 and 12 mOsm, and only
three washings were necessary (30 min each at 12000 rpm) to
obtain 'pale pink' ghosts. This modification allowed us to obtain
ghosts in the early afternoon, in time to perform the assay of
ATPase activity the same day.

The enzymatic activity was determined by measuring the
release of inorganic phosphorus (Pi) in an assay medium
containing: suspension of 'ghosts' (80-120 ug protein), 92 mM
Tris buffer pH 7.5, 5 mM $MgSO_4$, 5 mM KCl, 60 mM NaCl, and 0.1 mM
EDTA (Bonting, 1970). The reaction was started after
equilibration for 10 min at 37°C in a shaking bath, by the
addition of 4 mM ATP to a final volume of 2.2 ml, and stopped
after 15 min by the addition of 10% trichloroacetic acid. Na/K
ATPase activity was routinely obtained as previously described
(Luly and Verna, 1974) by subtracting the activity in the medium
lacking potassium (Mg^{++} ATPase) from the total activity.

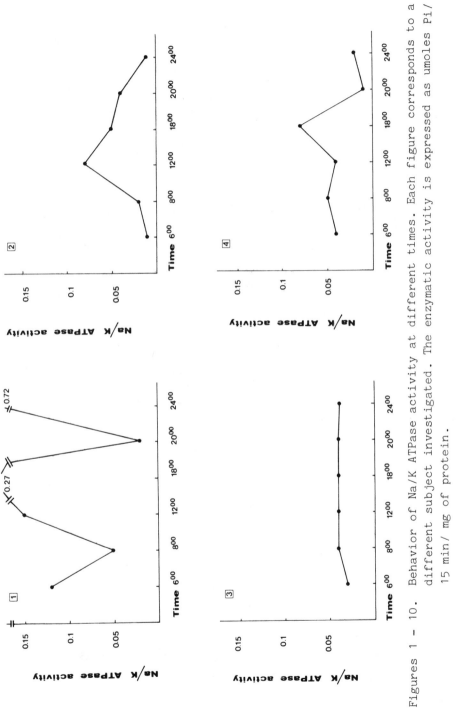

Figures 1 – 10. Behavior of Na/K ATPase activity at different times. Each figure corresponds to a different subject investigated. The enzymatic activity is expressed as umoles Pi/ 15 min/ mg of protein.

(continued)

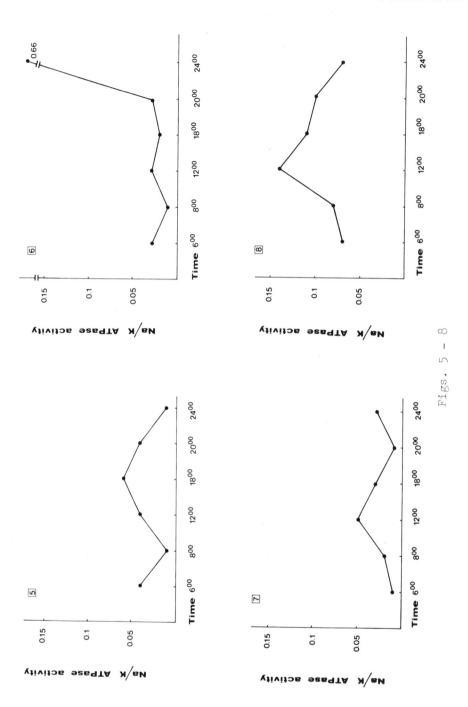

Figs. 5 – 8

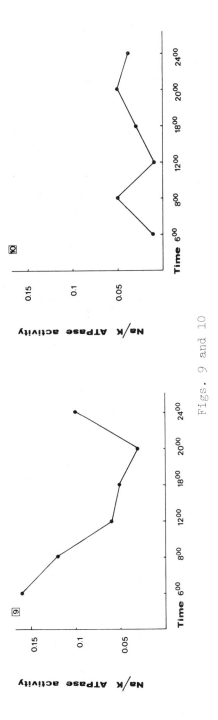

Figs. 9 and 10

TABLE I. Summary of Na/K ATPase activity at different times in
the different subjects investigated. For experimental details,
see text.

SUBJECT	TIME:					
	6:00	8:00	12:00	18:00	20:00	24:00
1	0.12	0.05	0.15	0.27	0.02	0.72
2	0.01	0.02	0.08	0.05	0.04	0.01
3	0.03	0.04	0.04	0.04	0.04	0.04
4	0.04	0.05	0.04	0.08	0.01	0.20
5	0.04	0.01	0.04	0.06	0.04	0.01
6	0.03	0.01	0.03	0.02	0.03	0.66
7	0.01	0.02	0.05	0.03	0.01	0.03
8	0.07	0.08	0.14	0.11	0.10	0.07
9	0.16	0.12	0.06	0.05	0.03	0.10
10	0.01	0.05	0.01	0.03	0.05	0.04

Table II. Individual rhythmometric analysis of Na/K ATPase
activity (umoles Pi/15 min/mg of protein). For experimental
details, see text.

SUB.	P	Mesor ±S.E.	Amplitude ±S.E.	Acrophase± S.E.
1	0.66	0.24±0.14	0.22±0.20	00.52 (04.25;21.21)
2	0.57	0.02±0.01	0.02±0.01	14.00 (16.44;11.16)
3	0.71	0.04±001	0.01±0.01	17.53 (22.34;11.52)
4	0.52	0.07±0.03	0.06±0.04	00.55 (03.36;22.14)
5	0.70	0.03±001	0.02±0.02	13.40 (17.36;09.44)
6	0.32	0.13±0.09	0.28±0.14	00.52 (02.44;23.00)
7	0.74	0.02±0.01	0.01±0.01	12.59 (17.13;08.45)
8	0.06	0.09±0.01	0.04±0.01	14.30 (15.24;13.36)
9	0.02	0.08±0.01	0.06±0.01	04.50 (05.38;04.02)
10	0.87	0.02±0.01	0.007±0.01	04.03 (12.54;19.52)

Experiments were run in triplicate. Protein content of the homogenates was assayed by the method of Lowry (Lowry et al., 1951). Time-qualified data were analyzed by the single cosinor method set up by Halberg (Halberg et al., 1972), a mathematical procedure which enables us to quantify the properties of circadian rhythms. The correlation between circadian parameters was statistically analyzed by the correlation coefficient.

RESULTS

In figures 1 to 10 the values of Na/K ATPase activity at different times and for each subject investigated are reported. These data are also summarized in Table I. At first glance appears that every subject (except one) shows several variations of the enzymatic activity during the 24-h span. Time-qualified data were therefore analyzed by the least square fit of a 24-h cosine function to derive the properties of an eventual circadian rhythm. Circadian parameters estimated were M.E.S.O.R. (rhythm-adjusted mean), amplitude (one-half of the extent of waveform oscillation), and acrophase (lag of the crest from a conventional time, here local midnight). The chronobiologic procedure documented a circadian rhythm for the Na/K ATPase activity in 8 out of 10 subjects.

Interestingly the recognition of chronobiologic data reveals that almost every subject is characterized by substantial differences in rhythmometric properties when compared to the others (Table II). In other words, almost all the subjects present a circadian rhythm; but the characteristics of the rhythm are completely different from one individual to another.

DISCUSSION

Our results bring evidence for a rhythmical modification of the enzymatic activity of the Na/K ATPase complex, leading to the consideration of different hypotheses: (a) is the modification due to a modulation by circulating factors? or (b) is it due to structural changes intrinsic to the plasma membrane itself and thus able to modify its physiologic functions?

It has been reported that many physiologic conditions as well as biochemical substances, mainly hormones, modulate the activity of the sodium pump (Tria et al., 1974; Verna and Muraro,

1983; Verna et al., 1984). It is also known that most hormones are delivered to target cells according to a chronobiologic behavior. We therefore should admit the possibility that the circadian rhythm of the enzyme is the physiological result of the equilibrium between these hormones.

According to the second hypothesis, we should consider the possibility that membrane constituents are continuously and rhythmically rearranged, with the result of a rhythmic modification of the enzymatic properties of the membrane itself.

The biological concept of a continual replacement, i.e., turnover, of cellular constituents has been extensively developed in recent years, indicating that cellular membranes are also continually turning over and that constancy of structure exists within a continual and dynamic process of replacement of both proteins and lipids. Despite the large quantity of methods used to clarify the mechanism of synthesis, assembly and turnover of membrane constituents, little is known yet, and a definitive answer must wait the development of more sophisticated methods for isolating and identifying membrane proteins.

Here we want to propose a new way to approach the study of plasma membrane physiology and function, by combining the two hypotheses, and advancing a new suggestive one: that the continuous replacement proper of the membrane structure, coupled with the action of circulating factors, might result in a modification of some plasma membrane activities with the features of a biological rhythm.

ACKNOWLEDGEMENTS

This work was supported by Universita' di Roma, progetto di ateneo "Fenomeni di Membrana". The authors also wish to thank Silvana Savini and Sergio Ferraro for their helpful technical assistance.

REFERENCES

Bonting, S.L. (1970). Sodium-potassium ATPase and cation
    transport. In: (E.E. Bittar Ed.) Membranes and ion
    transport, Vol I, pp. 257-363, Wiley Interscience
Halberg, F., Johnson, E.A., Nelson, W., Runge, W., Southern, R.B.
    (1972). Autorhythmometry: procedures for physiologic
    self measurements and their analysis. Physiology Teacher
    1:1.
Hanahan, D.J. and Ekholm, J.E (1974). The preparation of red cell
    ghosts (membranes). Meth. Enzym. 15, 168-172
Lowry, O.H., Rosebrough, N.J. Farr, A.L. and Randall, P.J.
    (1951). Protein measurement with the Folin phenol
    reagent. J. Biol. Chem. 193, 265-275.
Luly, P.,and Verna, R. (1974). Stimulation of Na/K ATPase of rat
    liver plasma membrane by aminoacids. Biochim. Biophys.
    Acta 367; 109-113
Tria, E., Luly, P., Tomasi, V., Trevisani, A. and Barnabei, O.
    (1974). Modulation by cAMP in vitro of liver plasma
    membrane Na/K ATPase and protein kinase. Biochim. Biophys.
    Acta 343, 297-306
Verna, R., Braquet, P., Diez, J. and Garay, P. (1984). Target
    cell responsiveness to peptide hormones, This Volume, page
Verna, R. and Muraro, R. (1983). Essential hypertension: Role of
    Na/K ATPase and relationship with the cyclic nucleotides
    system. Cell. Mol. Biol. 29 (1), 93-102

THE INVOLVEMENT OF cAMP IN THE HORMONAL REGULATION OF ION TRANSPORT

Ricardo Garay, Javier Diez, Roberto Verna,
Corinne Nazaret and Pierre Braquet *

INSERM U7/CNRS LA 318, Hôpital Necker, 161 rue
de Sevres, Paris 75015; France

* I.H.B. 17, av. R. Descartes, Le Plessis, France

INTRODUCTION

Ion metabolism is catalyzed by transport proteins of cell membranes. Generally speaking, one can differentiate between: i) the absorption and excretion of ions through epithelial tissues and ii) the exchange of ions between intra and extracellular compartments. It is now well established that transepithelial transport is under hormonal modulation. Recent results indicated that, even in non-epithelial cells, transmembrane ion movements are under hormonal control. This appears to involve cyclic AMP (cAMP) and other mediators (1,2,3,4).

The aim of our investigation is a better understanding of the interaction between cAMP and Na^+ transport proteins at the molecular level. Unfortunately, the knowledge of Na^+ metabolism at a molecular level is not as advanced as that of oxygen, sugar or lipid metabolism; this may be related to the fact that all known functional proteins of Na^+ metabolism are membrane transport proteins.

In avian red cells a bumetanide-sensitive Na^+,K^+ cotransport system catalyzes the simultaneous efflux or influx of both Na^+ and K^+ (1, 2). This trasport system is stimulated by a catecholamine-dependant adenylate cyclase system. The beta-adrenergic-dependent cotransport loop thus counterbalances the hyperkalemia induced by the intense muscular exercise of flight.

In mammals, the hyperkalemia of muscular exercise appears to be corrected by the beta-adrenergic stimulation of the ouabain-sensitive Na^+,K^+-pump of skeletal muscle (4).

One successful approach towards elucidating the molecular mechanism of Na^+ transport is the study of transport kinetics. This approach has recently permitted the characterization of different Na^+ transport systems in human red cells (5,6,7,8,9). On the other hand and in spite of the functional absence of beta-adrenergically stimulated adenylate-cyclase (10), human red cell membranes contain cAMP-dependant protein kinases (11). Thus it was of interest to see whether Na^+ and K^+ transport systems of human red cells might respond to cAMP independently of catecholamine receptor interactions in these cells. In addition, the effect of other hormonal intermediates, intracellular Ca^{2+} and cyclic GMP (cGMP) on ion transport was examined.

Na^+ AND K^+ TRANSPORT SYSTEMS IN HUMAN RED CELL MEMBRANES

The effect of different pharmacological agents on Na^+ fluxes, the kinetic properties of these fluxes and the comparative analysis of erythrocyte cation transport in various animal species have allowed the characterization of the different Na^+ and K^+ transport systems existing in human red cell membranes (Figure 1 and ref. 5,6,7,8,).

Most of the carrier mediated Na^+ and K^+ fluxes across human red cell membranes are catalyzed by the ouabain-sensitive Na^+,K^+-pump. This system catalyzes the exchange of intracellular Na^+ for extracellular K^+ coupled to the hydrolysis of ATP, thus generating electrochemical gradients of Na^+ and K^+ across the cell membrane. The stationary intracellular Na^+ concentration results from the balance between the Na^+ efflux catalyzed by the pump and the net Na^+ influx through the lipid bilayer and the anion carrier. It is now well documented that human red cell membranes contain other minor Na^+ transport systems independent of the Na^+,K^+-pump. One of the best characterized is the chloride-dependant Na^+,K^+ cotransport system (6,7). This system catalyzes coupled movements of Na^+,K^+ and perhaps Cl^- in inward and outward directions across the cell membrane. Another well characterized system is the one:to:one Na^+, Na^+ countertransport, which catalyzes very small fluxes in human red cells (8). Li^+ is used as a Na^+ analogue to measure this system as a Na^+, Li^+ countertransport.

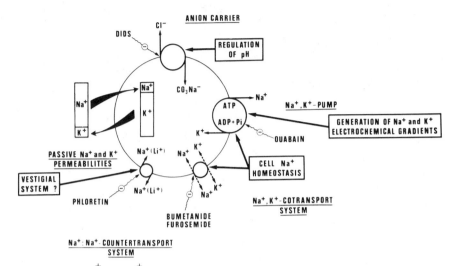

Fig. 1. Na^+ and K^+ transport across human red cell membranes

In addition to the transport systems described in Fig. 1, an internal Ca^{2+}-dependent K^+ channel ("Gardos effect") (12) and a NEM-stimulated Chloride-dependent K^+ transport system exist in human erythrocytes.

The physiological role of the erythrocyte Na^+ and K^+ transport systems

Under physiological conditions, the erythrocyte Na^+ concentration is about 15 times lower than the external Na^+ concentration and the intracellular K^+ concentration is about 30 times higher than the external K^+ concentration.
In human red cells, the membrane potential is about -10 mV and thus the pump-generated electrochemical energy is around 2000 cal/mol of Na^+ or K^+. Taking into account that the external Na^+ and internal K^+ concentrations are more than 100 times higher than the internal ATP concentration, a simple calculation indicates that the energy stored into electrochemical gradients is at least 10 times greater then that stored as ATP. In other non-erythrocyte cells this stored electrochemical energy is used for the

absorption of metabolic substrates, the extrusion of catabolites, the action potential and many other vital membrane functions. Each one of these functions is catalyzed by genetically coded transport systems, the gradient dissipators. The Na^+, Na^+ countertransport system of human erythrocytes may be a dissipator for more complex or more relevant transport functions in other cells.

Under basal conditions, the Na^+, K^+ cotransport system is near equilibrium, i.e.: the inward and outward cotransport fluxes show similar magnitudes and thus no net cation fluxes are observed. Any change in internal Na^+ or external K^+ concentration is corrected by a net cotransport flux in the opposite direction of the ionic perturbation. Thus, the cotransport system seems to contribute to the regulation of cell Na^+ and plasma K^+ concentrations.

INCORPORATION OF cAMP INTO HUMAN RED CELLS

A specific membrane transport system for the net extrusion of cAMP has been described in a variety of cells of different animal species including pigeon and rat erythrocytes (13). We have found that this system also exists in human red cells (3). Thus, the incubation of human erythrocytes in a medium containing 2 mM cAMP allows the incorporation of cAMP into the cells up to 50-70 micromoles/l. cells in 4-10 minutes (see ref. 3 for further details).

THE EFFECT OF EXOGENOUS cAMP ON ION TRANSPORT

Human erythrocytes were incubated in Mg^{2+} sucrose medium containing different concentrations of cAMP (see ref.3 for details). The fluxes catalyzed by the Na^+ pump were taken as the K^+-stimulated, ouabain-sensitive Na^+ efflux. The ouabain-resistant bumetanide-sensitive fraction of the Na^+ and K^+ effluxes was equated as the outward Na^+, K^+ cotransport fluxes.

The effect of cAMP on the erythrocyte Na^+, K^+ pump

Figure 2 shows in dose-response curve, the effect of exogenous cAMP on the ouabain-sensitive Na^+ efflux in erythrocytes from two healthy subjects. MIX (1-methyl-3-isobutylxanthine) is an inhibitor of phosphodiesterase.

Exogenous cAMP showed almost no effect on the fluxes catalyzed by the Na^+, K^+. In human red cells (Figure 2).

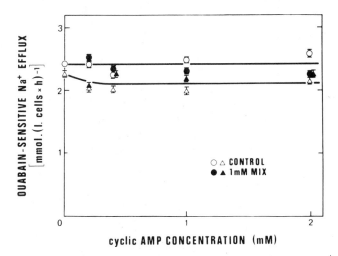

Fig. 2. Absence of effect of cAMP on the erythrocyte Na$^+$,K$^+$-pump. The red cells were drawn from two healthy donors.

The effect of cAMP on the erythrocyte Na$^+$ K$^+$cotransport system.

In contrast to the Na$^+$,K$^+$-pump, outward Na$^+$,K$^+$ cotransport was affected by the presence of the cAMP in the incubation medium. A dose-response curve showed marked inter-individual variations between four different donors (Fig. 3). Thus the external cAMP concentration giving 50% cotransport inhibition varied from 0.1 to 5 mM. This corresponds to an upper limit of 3 to 150 micromoles/l. cells of intracellular cAMP (see ref. 3 for details). This seemed sufficient to saturate the cAMP-dependent protein kinase, which shows 50% activation for cAMP concentrations ranging from 0.1 to 1 micromol (11). However, the actual cAMP concentration is probably less then the one estimated above, because a portion of intracellular cAMP is degraded by internal phosphodiesterases.

THE EFFECT OF cGMP AND Ca^{2+}ON ERYTHROCYTE CATION TRANSPORT

cGMP only minimally inhibited the fluxes catalyzed by the Na$^+$,K$^+$-pump. Figure 4 shows a dose-response curve in erythrocytes from two healthy donors. Interestingly, in one of the donors the inhibition is significantly enhanced by inhibition of phosphodiesterase with MIX.

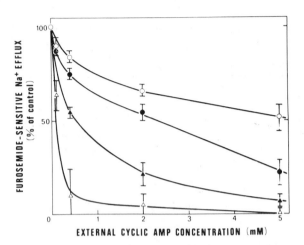

Fig. 3. Inhibition of outward Na^+,K^+ cotransport fluxes
by exogenous cAMP in human red cells.

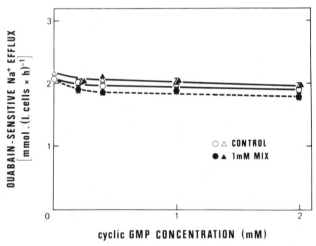

Fig. 4. The effect of cGMP on the erythrocyte Na^+,K^+-pump.

Outward Na^+,K^+ cotransport fluxes were only slightly inhibited by cGMP (see ref. 3). Conversely, if the Ca^{2+} ionophore A23187 is added to a 1 mM Ca^{2+} solution, the cotransport system is strongly inhibited (see ref. 3)

DISCUSSION

The Na^+,K^+-pump and Na^+,K^+ cotransport system are widely distributed among several kinds of cells. The pump activity generates low concentrations of both internal Na^+ and esternal K^+. The cotransport system counterbalances any variation in these two parameters.

In several cells cAMP may activate the pump and inhibit the cotransport system. For instance cAMP activate the Na^+,K^+ pump of skeletal muscle (4) and elicited peritoneal mouse macrophages (J. Diez, R. Verna, at al., unpublished results). In addition cAMP inhibits the Na^+,K^+ cotransport system of human erythrocytes and mouse macrophages. This allows us to postulate the model of Fig. 5 in which cAMP activates the gradient generator and shut off the gradient regulator. The net result may be an increase in the transmembrane Na^+ and K^+ electrochemical gradients. If these gradients are strongly perturbed, the effect of cAMP may be to restore the altered equilibrium. However, care must be taken with the extrapolation of this model to any kind of cell because in some cell types such as hepatocytes, cAMP may inhibit the Na^+,K^+ pump and in avian erythrocytes may activate the Na^+,K^+ cotransport system.

In addition to the effect of cAMP other mediators may affect ion transport. Internal Ca^{2+} may inhibit both, the Na^+,K^+-pump (14) and the Na^+,K^+ cotransport system (3). Thus a number of circulating hormones may act on Na^+ and K^+ transport via cAMP and Ca^{2+} (Figure 5)

In conclusion, our results further suggest the existence in man of a hormonal network for the regulation of Na^+ and K^+ electrochemical gradients throught such intermediates as cAMP and Ca^{2+}. Recent results from our laboratory indicate that such regulation may be perturbed in some patients with essential hypertension. In these subjects, the Na^+,K^+ cotransport system or the Na^+,K^+-pump are not able to adequately regulate an increase' in cell Na^+ concentration.

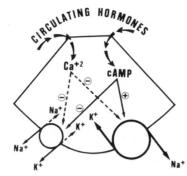

Fig. 5. The hormonal network model for the regulation of Na[+] and K[+] electrochemical gradients across cell membranes.

REFERENCES

1. T.J. McManus and W.F. Schmitt, in: Membrane Transport Processes (Hoffman, J. F., ed.) Raven Press 1:79 (1978).

2. S.L. Alper. H.C. Palfrey, S.A. De Riemer and P. Greengard, J. Biol. Chem. 255:11029 (1980).

3. R.P. Garay, Bioch. Bioph. Acta 688:786 (1982).

4. T. Clausen, Acta Med. Scand. suppl; 672:111 (1983);

5. R.P. Garay and P.J. Garrahan, J. Physiol. (London) 231:297 (1973).

6. J.S. Wiley and R.A. Cooper. J. Clin. Invest. 53:745 (1974).

7. R.P. Garay, N. Adragna, M. Canessa and D.C. Tosteson, J. Membrane Biol. 62:169 (1981)

8. B. Sarkadi, J.F. Alfinoff, R.B. Gunn and D.C. Tosteson, J.Gen. Physiol. 72:249 (1978).

9. J. Funder, D.C. Tosteson and J. Wieth, J. Gen Physiol. 71:721 (1977).

10. H. Rasmussen, W. Lake and J.E. Allen, Bioch.Bioph. Acta. 411:63 (1975).

11. B. F. Grant, T. Breithaut and E.B. Cunningham, J. Biol. Chem. 254:5726 (1979).

12. V. Lew and L. Beauge', Membrane transport in biology 2:81 (1979)

13. G. Wiemer, U. Hellwich, A. Wellstein, J. Dietz, M. Hellwich and D. Palm, Naunyn-Schmiedeberg's Arch Pharmacol. 321:239 (1982).

14. B. Sarkadi, I. Szasz, A. Gerloczy and G. Gardos, Bioch. Biophys. Acta 464:93 (1977).

EFFECTS OF GLUCOCORTICOIDS ON HORMONAL ACTIVATION OF ADENYLATE CYCLASE

G. Tolone, L. Bonasera and P. Lo Presti

Istituto di Patologia Generale
Università di Palermo
Corso Tukory 211, Palermo, Italy

INTRODUCTION

It has been reported (1-4) that several eukaryotic cells respond to prostaglandins, catecholamines or beta-adrenergic agonists with a greater increase in cellular cyclic AMP when exposed in vitro to hydrocortisone or dexamethasone. Although several aspects of this synergism have been investigated, very little is known yet about the mechanisms underlying its action. Direct binding studies (4,5) using tritiated dihydroalprenolol have suggested that glucocorticoid potentiation probably results from events occurring beyond the interaction of the ligand to its receptor. An attractive hypothesis, which mechanism is involved in the hormonal stimulation of adenylate cyclase activity, is that corticosteroids favour the "opening" of the guanyl nucleotide site of adenylate cyclase thus increasing its activation by catecholamines and other hormones. Alternatively, since it has been shown (7) that in many cell types phospholipid methylation in a local area in the membrane changes the miroenvironment in such a way as to facilitate the lateral mobility of the receptor complex, it can also be hypothized that glucocorticoids favour receptor-adenylate cyclase coupling by increasing the activity of membrane methyltransferases. To test both hypotheses, we have investigated whether hydrocortisone or dexamethasone, in pharmacologically attainable concentrations,

enhance isoproterenol-induced release of bound GDP from
erythrocyte membrane, and/or if they increase isoproterenol-
induced phospholipid methylation in rat reticulocyte ghosts.

METHODS

Measurement of isoproterenol-induced release of
----
membrane-bound nucleotide.

Turkey erythrocyte membranes were prepared according to Cassel
and Selinger (8) and preincubated for 2 min at 30°C in a medium
containing 6mM $MgCl_2$, 0,3 mM ATP, 12 mM creatine phosphate, 50
units/ml of creatine phosphokinase, 2 mM 2-mercaptoethanol, 0,2
mM ethylene glycol-bis (beta-aminoethyl ether)-N,N'-tetraacetic
acid (EGTA), 20 uM dl-isoproterenol and 50 mM
3-(N-morpholino)-propanesulfonic acid (Mops), (ph 7,4). The
binding of tritiated nucleotide was initiated by the addition of
0,25 uM ^{3}H-GTP (4,000 cpm/pmol), and it was terminated after a 2
min incubation at 30°C by the addition of 100 uM unlabelled GTP
and 20 uM propranolol. The reaction mixture was cooled, and the
membranes were washed three times by centrifugation and
resuspended in cold 10 mM Mops (pH 7,4) containing 2mM
2-mercaptoethanol. Aliquots of 1 vol of membrane suspension (2
mg/ml) were then added to 4 vol of medium containing 6 mM $MgCl_2$,
2 mM 2-mercaptoethanol, 0,2 mM EGTA, 50 mM Mops (pH 7,4) and 10
uM hydrocortisone, dexamethasone, progesterone or testosterone.
The control system did not receive steroids. After 5 min
incubation at 37°C, the reaction was started by the addition of
0,1 mM guanosine 5'-(beta, gamma-imino) triphosphate (Gpp(NH)p)
and 50 uM di-isoproterenol, and was stopped 3 min later by the
transfer of 0,5 ml aliquots into 1,5 ml of ice cold Mops buffer.
The resultant mixture was centrifuged (12,000 g for 10 min) and
the supernatant was assayed for radioactivity in a scintillation
spectrometer.

Determination of isoproterenol-induced phospholipid
----
methylation.

Rat reticulocyte ghosts were prepared according to Hirata et al.

(9) and resuspended in 50 mM Tris-glycylglycine buffer (ph 8,0)
made 5 mM $MgCl_2$. After centrifugation, the packed ghosts (0,5 ml)
were suspended in 1,5 ml of the same buffer containing 0,4 mM
S-adenosyl-methyl-^{3}H-methionine (800 dpm-pmol). After incubation
for approximately 15 hr at 4°C, 10 ml of 0,9% NaCl were added to
the mixture and then centrifugated. The pelled was resuspended in
100 mM Tris-glycylglycine buffer pH 8,0) made 10 mM $MgCl_2$. After
10 min preincubation at 37°C, aliquots (100 ul) of the ghost
supsension were transferred to a test tube containing an equal
volume of hydrocortisone or dexamethasone (10 uM final
concentration), and the reaction was started by the addition of
200 ul of dl-isoproterenol (0,1 mM final concentration). After a
suitable time interval (usually 30 min), 200 ul samples were
pipetted into tubes containing 2 ml of chloroform/methanol/2 M
HCl (6:3:1) for phospholipid extraction. After vortex mixing, the
aqueous phase was removed and the chloroform phase was washed
twice with 0,1 M KCl in 50% methanol, dried at room temperature,
dissolved in Triton X-100/toluene-base scintillator and assayed
for radioactivity.

RESULTS AND DISCUSSION

As indicated in Table 1, hydrocortisone and dexamethasone, in
a pharmacologically attainable concentration (0,01 mM),
significantly enhance isoproterenol-induced release of tritiated
nucleotide, mostly GDP (result not shown), from erythrocyte
membranes pretreated with ^{3}H-GTP. This effect seems to be
specific for glucorticoids since progesterone and testosterone
are both inactive. By contrast, neither hydrocortisone nor
dexamethasone affect isoproterenol-induced phospholipid
methylation in reticulocyte ghosts (Table 2). This finding, which
argues against an involvement of membrane methyl-transferases in
the synergism between corticosteroids and those hormones whose
effects are mediated through adenylate cyclase activation,
apparently disagrees with other reports (10). Such a discrepancy
probably reflects differences in the experimental models;
reduction of phospholipid methylation by dexamethasone has in
fact been observed exclusively in nucleated cells (mastocytes),
and only after exposition of these cells to the steroid for a
long time interval (12-24 hr). Since the potentiating effect of
glucorticoids shows no latency, it seems reasonable to presume

Table 1. Effect of several steroids on isoproterenol-induced
release of tritiated nucleotides from turkey erythrocyte
membranes pretreated with ^{3}H-GTP *

```
----------------------------------------------------------------
                        3
                         H-nucleotide released
Steroid                 (pmol/mg protein)
----------------------------------------------------------------

None                    0.95 ± 0.14
Dexamethasone           2.14 ± 0.21 **
Hydrocortisone          2.02 ± 0.22 **
Progesterone            0.87 ± 0.09
Testosterone            1.04 ± 0.17
----------------------------------------------------------------
```

* Membranes pretreated with 0.25 uM ^{3}H-GTP in a medium containing
20 uM isoproterenol were washed and assayed for nucleotide
release in the presence of 0.1 mM Gpp(NH)p and 10 uM of the
indicated steroids. Results are expressed as means ± SEM (5
experiments).

** Differences from control value significant at P. 0.05.

Table 2. Effect of hydrocortisone and dexamethasone on
isoproterenol-induced phospholipid methylation in reticulocyte
ghosts ***.

```
        -------------------------------------------------
                            3
                             H-methyl incorporation
        Ghosts exposed to  (pmol/mg protein)
        -------------------------------------------------
        Buffer alone            28 ± 5.0
        Buffer + Hydrocortisone 31 ± 6.6
        Buffer + Dexamethasone  26 ± 3.8
        -------------------------------------------------
```

*** Reticulocyte ghosts pretreated with 0.4 mM
S-adenosyl-L-methyl- ^{3}H methionine and exposed (or not) to 10 uM
hydrocortisone or dexamethasone were assayed for phospholipid
methylation in the presence of 0.1 mM isoproterenol. Ghosts were
collected for lipid extraction and radioactivity assay 30 min
after isoproterenol challenge. Results are expressed as means ±
SEM (3 experiments).

that it is independent of modification of membrane fluidity, a
statement which is further supported by the finding of Lewis et
al. (11) that anti-infiammatory steroids do not alter the
transition temperature of dipalmitoyl lecithi liposomes.

Our results point rather to the guanyl nucleotides binding
site of adenylate cyclase as a target for glucocorticoids. Since
the binding of guanyl nucleotide analogs to the adenylate cyclase
complex is a cation-dependent phenomenon (6), it can be supposed
that steroids facilitate catecholamine-induced release of bound
GDP by decreasing its affinity for the regulatory site of the
enzyme through an effect on cation influx across the membrane.
Experiments are in progress to verify this hypotesis.

REFERENCES

1. C.W.Parker, M.G.Huber, and M.L.Baumann, Alterations in cyclic
   AMP metabolism in human bronchial asthma. III Leukocyte
   responses to steroids, J. Clin. Invest. 52:1324 (1973).
2. J. Mendelsohn, M.M.Mutler, and R.F. Boone, Enhanced effects of
   prostaglandin E1 and dibutyryl cyclic AMP on human
   lymphocytes in the presence of cortisol, J. Clin. Invest.
   52:2129 (1983).
3. S.J. Foster and J.P. Perkins, Glucocorticoids increase the
   responsiveness of cells in culture to prostaglandin E1,
   Proc. Natl. Sci. USA. 74:4816 (1977).
4. G. Tolone, L. Bonasera, and R. Sajeva, Hydrocortisone
   increases the responsiveness of mast cells to beta-adrenergic
   agonists by an action distal to the beta-adrenoreceptors,
   Br. J. Exp. Path. 60:269 (1979).
5. D.L. Marquardt and S.I. Wasserman, Modulation of rat serosal
   mast cell biochemistry by in vivo dexamethasone
   administration, J. Immunol. 131:934 (1983).
6. D. Cassel and Z.Selinger, Mechanism of adenylate cyclase
   activation through the beta-adrenergic receptor:
   catecholamine-induced displacement of bound GDP by GTP,
   Proc. Natl. Acad. Sci. USA. 75:4155 (1978).
7. F. Hirata and J. Axelrod, Phospholipid methylation and
   biological signal transmission, Science, 209:1082 (1980).

8. D. Cassel and Z. Selinger, Mechanism of adenylate cyclase
   activation by cholera toxin: inhibition of GTP hydrolysis
   at the regulatory site, Proc.Natl.Acad. Sci. USA. 74:3307
   (1977).

9. F. Hirata, W.J. Strittmatter, and J. Axelrod, Beta-adrenergic
   receptor agonists increase phospholipid methylation,
   membrane fluidity, and beta-adrenergic receptor-adenylate
   cyclase coupling, Proc. Natl. Acad. Sci. USA. 76:368 (1979).

10. M. Daeron, A.R. Sterk, and T. Ishizaka, Biochemical analysis
    of glucocorticoid-induced inhibition of IgE-mediated
    histamine release from mouse mast cells, J. Immunol. 129:1212
    (1982).

11. G.P. Lewis, P.J.Piper, and C. Vigo, Mechanism of action of
    anti-inflammatory steroids on membrane fluidity and
    phospholipase activity. Br. J. Pharmac. 67: 453P (1979).

THE ROUND TRIP OF THYROGLOBULIN WITHIN THE THYROID CELL:

CELLULAR RECEPTOR(S) AND MOLECULAR SIGNALS FOR THE RECOGNITION

Silvestro Formisano*, Leonard D. Kohn** and
Eduardo Consiglio*

*Centro di Endocrinologia ed oncologia sperimentale
del C.N.R., c/o Istituto di Patologia Cellulare e
Molecolare L. Califano, II Facolta' di Medicina e
Chirurgia, Universita' degli Studi di Napoli, 80131
Napoli, Italy
**Section of Biochemistry of Cell Regulation
Laboratory of Biochemical Pharmacology
National Institutes of Arthritis, Diabetes, and
Digestive and Kidney Diseases
National Institutes of Health
Bethesda, Maryland 20205

Inside the thyroid gland, composed of follicular structures, a complex pattern of molecular events occurs for the transport of the thyroglobulin (Tg)*.

Thyroglobulin is the major glycoprotein synthesized by the thyroid cells. It is a glycoprotein having a large molecular weight (660,000 daltons) which represents the molecular site for the thyroid hormone synthesis (1). The synthesis of Tg occurs on the membranes of the rough endoplasmic reticulum (RER) of the thyroid cells; it is then transported to the apical border of the cell were it is secreted and stored within the follicular lumen. The degradation of the Tg depends on highly selective pino-endocytotic process which occurs through a series of specific interactions with the plasma membranes and the lysosomal systems, where Tg is completely hydrolyzed and the thyroid hormones T_3 and T_4 secreted into the blood stream.

---

* The abbreviations used are: Tg, thyroglobulin, T_3, triiodothyronine; T_4, tetraiodothyronine; RER, rough endoplasmic reticulum; Gluc, glucose; Man, mannose; GlucNac, N-acetylglucosamine; Gal, galactose.

This chapter is in the nature of a progress report on recent work done to characterize thyroglobulin–thyroid cell membrane interactions and its molecular post-translational modifications which are involved in the biosynthetic and biodegradation mechanism of the thyroglobulin.

The major post-translational events of the Tg molecule are: glycosilation, iodination and phosphorilation.

SYNTHESIS OF THE CARBOHYDRATE CHAINS   AND FUNCTIONAL ROLE OF THE EXPOSED SUGAR ON THYROGLOBULIN

A) SYNTHESIS OF THE CARBOHYDRATE CHAINS: The first post-translational modification (event) which occurs on the Tg polypeptide chain (12 S subunit) is the synthesis of the carbohydrate chains. Such synthesis is similar to those resolved with other glycoproteins but is less well documented. In figure 1 we report the structure available on the A and B carbohydrate chains of thyroglobulin. Two major distinct carbohydrate chains

GLYCOPEPTIDE A FROM HUMAN AND CALF THYROGLOBULIN

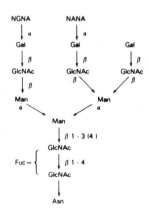

GLYCOPEPTIDE B FROM PORCINE THYROGLOBULIN

Figure 1: Structure of carbohydrate chains.

are present on the Tg molecules, a series of different polymannose chains, termed the A unit, and more complex carbohydrate chains, termed the B unit (2-6).
Both chains are synthesized by a single polysaccharide which is attached to the amide nitrogen of the asparagine group and is in the sequence Asn-X-Ser(Thr) (6).

The single polysaccharide precursor $(Gluc_3-Man_9-GlucNac_2)$ is sequentially synthesized via dolichol-phosphate cycle, the functional lipid moiety (6). The oligosaccharide precursor is then degradaded in to a simpler structure, several steps being involved in the enzymatic hydrolysis of the carbohydrate (6).First, glucose residues are removed, followed sequentially by the trimming of several mannoses and the addition of a new complement of terminal sugars (GlucNac, Gal, Sialic acid and Fucose units) via the activity of several different glycosyltransferases. Terminal glycosylation and at least some mannose remodeling occurs after the transfer of newly synthesized protein from the ER to Golgi apparatus.

Galactosyl-transferase and sialyl-transferase are the "late acting" enzymes which are involved in the terminal glycosylation of the complex units. These steps are likely located in the exocytotic vesicles or in the apical membranes.

B) THE N-ACETYLGLUCOSAMINE AND SIALIC ACID ARE SIGNALS IN THE RECOGNITION OF THE SPECIFIC BINDING SITE TO CELL MEMBRANE

As we can observe from figure 1 the removal of the sialic acid and galactose residues from the B carbohydrate chain exposes N-acetylglucosamine residues. The effects of such exposure on the binding of thyroglobulin to thyroid membranes and its possible physiological role has been the subject of several studies (13, 14).
The increase in binding in ^{125}I-labeled asialothyroglobulin preparations after beta-galactosidase treatment, as demonstrated in the table, indicated that N-acetylglucosamine residues on the B carbohydrate chain of thyroglobulin were important in the receptor recognition process.

As reported in the table, the absence of any effect of beta-galactosidase on labeled native thyroglobulin indicates that the activity is related to the exoglycosidase action of beta-galactosidase preparation rather than to a combinant endoglycosidase or protease activity.
It was also shown (14) that there is an inverse relationship between exposed N-acetylglucosamine residues and the iodo-tyrosine content of the thyroglobulin. Such findings have suggested that thyroglobulin molecules with exposed N-acetylglucosamine residues would be protected from degradation by binding to the thyroid membranes, but would be exposed to processes which result in the iodination of tyrosyl residues of thyroglobulin.

Table I.   Ability of Alfa-galactosidase or Beta-galactosidase
           Preparations to Modify Thyroglobulin or Asialothyro-
           globulin Binding to Thyroid Membrane Preparations.

Thyroglobulin preparation $	Enzyme pretreatment	Binding at pH 6.0 and 0°C for 90 min	
		cpm bound	% of control
^{125}I-labeled asialo-thyroglobulin	None	8,950	100
	Beta-Galactosidase	58,500	654
	Alfa-Galactosidase	12,500	140
	Almond emulsion	2,000	22
^{125}I-labeled thyroglobulin	None	2,500	100
	Beta-Galactosidase	2,400	96
	Alfa-Galactosidase	2,700	108
	Almond emulsion	2,800	112

$ See Material and Methods of ref. 13, 14 for pretreatment
details.

        Ashwell and his colleagues (16) have defined a liver
membrane hepatic binding protein capable of recognizing asialo-
glycoprotein after sialic acid removal and exposure of the
penultimate galactosyl residues. The galactose binding protein
has been associated with the internalization and degradation of
the protein. The sialic residues on a glycoprotein thus actually
prevent a protein from binding to the membranes and therefore
from being degraded. The GlcNac binding protein on thyroid
membranes can be considered an analog of the hepatic binding
protein,   in   fact   immunologic   cross-reactivity   has   been
demonstrated (13, 14). In the case of the thyroglobulin the
GlcNac binding is associated with the vectorial transport of Tg
to follicles since the sequential maturation of the B CHO chain,
i.e. galactose and sialic residues, are associated with the
iodination and release of Tg into the follicular lumen (13, 14,
15, 17). The level of iodination is already been accepted as a
means of Tg maturation. It has been suggested that as the sialic
acid residues covers GlcNac-Gal residues and as iodine covers
tyrosine residues, membrane binding is terminated and the
iodinated Tg molecule enters the follicular lumen were it is

stored until called for by the specific activation at the
pinocytotic secretion mechanism by TSH. The Tg receptor thus
seems to be operating as the Ashwell receptor but in an
opposite directional mode: in to out rather out to in. But in
both cases the role of the sialation is the same, i.e. to
prevent the membrane interaction and protein degradation until
specific mechanisms are unvoked.

These data have been used to reexplain the original data of
Salvatore and colleagues (12). The low iodinated molecules with
exposed GlcNac-Gal residues would be bound to membrane and be in
the midst of the maturation process. Since the pinocytotic
process involves only a small quantity of membrane structure,
these would be protected while the pinocytotic process
"selectively" engulfs high iodine, high sialic Tg molecules.

ROLE OF IODINE ON THE STRUCTURE AND ON THE TRANSPORT OF THE
THYROGLOBULIN IN THE FOLLICLES: The iodination is the specific
processing modification which occurs on the Tg, since the iodine
is necessary for the synthesis of the thyroid hormones,
triiodothyronine and tetraiodothyronine (7-9). In fact, the
release of the thyroid hormones into the blood stream depends
on: (i) the movement of the thyroglobulin molecules (which are
randomly iodinated) from the follicular lumen into the cell
compartment through a selective pino-endocytotic process which
is TSH and c-AMP mediated; (ii) the intracellular hydrolytic
cleavage which occurs within the lysosomal system; (iii) the
transfer of the hormones (T_3 and T_4) from the cell to the
blood stream.

Salvatore and colleagues (10-12) have shown that such
secretory process (in vitro and in vivo experiments) is
selective in that there is preferential reabsorption of more
heavily iodinated pulse-labeled thyroglobulin like iodoproteins
from follicular structures and a protection of newly formed,
poorly iodinated thyroglobulin molecules from the endocytotic
process.

One possibility to explain selective segregation of
macromolecules is by means of specific membrane or receptor
interactions. The hypothesis that these were the basis for the
selective degradation of high iodine molecules evolved from
studies which adapted techniques used in studies of
hormone-receptor interactions. The conclusion reached, using
binding experiments, was that the cell thyroid membranes have
receptors which are capable of the specific recognition of
thyroglobulin and that this recognition process involves: the
extent of iodination of the tyrosine and, therefore, the
physical state of the thyroglobulin.

Since thyroglobulin with different iodine content bind to
the thyroid membranes with different extent, the role of tyrosyl

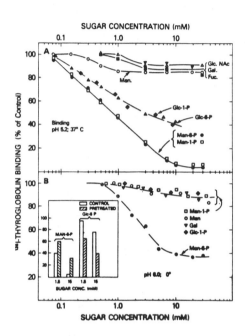

Figure 2: The binding of ^{125}I-labeled thyroglobulin and its derivative to thyroid membranes at room temperature, pH 7.0
A: (●) unmodified thyroglobulin; (■) 10:1 nitro-thyroglobulin; (▲) 10:1 0-acetyl-thyroglobulin.
B: solid curve: 10:1 0-acetyl-thyroglobulin; broken curve: 10:1 0-acetyl-thyroglobulin after removal of the acetyl groups with 1M hydroxylamine, pH 7.5. The concentration of the membranes was 50 ug.

residues has been examined. Tyrosyl residues have been modified by reacting with tetranitromethane or with N-acetyl-imidazole. Fig. 2 A shows the binding of thyroglobulin with nitrate or acetylated tyrosine groups to the thyroid plasma membranes. The results clearly indicate that the modification of the tyrosyl residues interferes with the membrane interaction. The key role of the tyrosyl residues in the binding to the thyroid membranes is more readily demonstrated with the 0-acetyltyrosine derivative by N-acetyl imidazole of thyroglobulin, as reported and as also demonstrated in Fig. 2 A. In fact, the modification of only 70% of the tyrosyl residues by 0-acetylation dramatically reduced the ability of thyroglobulin to bind to thyroid membranes.

This decreased binding could be reversed by removing the 0-acetyl group with hydroxylamine as shown in Fig. 2 B. As shown the binding of hydroxylamine-treated thyroglobulin is greater than the modified protein since N-acetyl-imidazole will also acetylate lysine residues, lysine coupled acetyl groups are not removed with hydroxylamine, and modification of lysine residues can enhance binding.

D) PHOSPHATE RESIDUES ON THYROGLOBULIN ARE SPECIFIC MEMBRANE RECOGNITION EVENTS. During the studies on thyroglobulin binding to thyroid membranes at pH 5 or pH 6 it was noted that mannose-6-phosphate was excellent inhibitor of thyroglobulin interaction with thyroid membranes as reported in Figure 3. At pH 6.0 (Fig. 3 panel B), the mannose-6- phosphate is specific compared to mannose-1-phosphate or glucose-6-phosphate. The poor specificity of mannose-6-phosphate at pH 5 (Fig. 3 panel A) is explained when pre-incubation of thyroglobulin with pH 5 membrane extracts showed that there were isomerases or epimerases which decreased mannose-6-phosphate inhibition and increased glucose-6-phosphate or inhibition (Fig. 3 B insert). The interpretation of these data suggested the possibility that sugar phosphate might be present on thyroglobulin and that a sugar phosphate recognition system might exist on thyroid cells and be important in lysosomal targeting.

The presence of phosphate on thyroglobulin has now been established by Consiglio and colleagues (18). 10 moles of phosphate per mole of thyroglobulin, was found. The phosphate is covalentlhy attached to thyroglobulin and appears to be of two types. One fraction is located on endo-H sensitive oligosaccharides of Tg, i.e., the A-chain; the second group is on peptides also containing B carbohydrate chains. Nuclear magnetic resonance studies indicate the latter group may be tyrosine phosphate. The role of the phosphate residues is unclear but analogies to other system would predict their importance in vectorial movement from E.R. to Golgi (tyrosine posphate) and from E.R. to lysosomes (sugar posphate) (19) those remain to be evaluated.

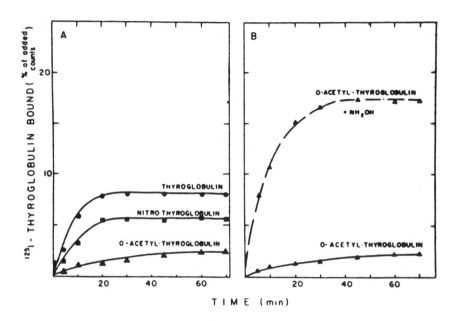

Figure 3: (A) Inhibition of ^{125}I-labeled thyroglobulin binding to bovine thyroid membranes by the noted concentrations of mannose-6-phosphate (Man-6-P), mannose-1-phosphate (Man-1-P), glucose-6-phosphate (Glc-6-P) at pH 5.2 and 37°C. Incubations contained 300,000 cpm labeled thyroglobulin and 40 ug membrane protein. Binding in the absence of any sugar phosphate (100% of control) was 132,000 cpm; control values in the absence of membranes as with unlabeled thyroglobulin were 5% or less these values. (B) Inhibition of labeled thyroglobulin binding to bovine membranes at pH 6.0 and 0°C. Conditions were otherwise the same as in (A); The 100% value in this case was 18,000 cpm bound; mannose is designated (Man). The insert depicts binding at pH 6.0 and 0°C after thyroglobulin preparations were preincubated for 1 hr with either (i) a membrane supernatant preparated by extracting membranes at pH 5.0 and 37°C (preincubation) or (ii) the same membrane supernatant after heating for 10 min. at 95°C (control). As noted mannose-6-phosphate inhibition is less whereas the glucose-6-phosphate inhibition is enhanced.

A CURRENT WORKING SCHEME OF THYROGLOBULIN EXOCYTOSIS AND ENDOCYTOSIS.

Current data (20) indicate that the polysomal thyroglobulin synthetic product is the polypeptide portion of the 12S, 330,000 molecular weight, subunit of the classically defined 19S thyroglobulin moiety isolated from the follicular lumen (Figure 4).

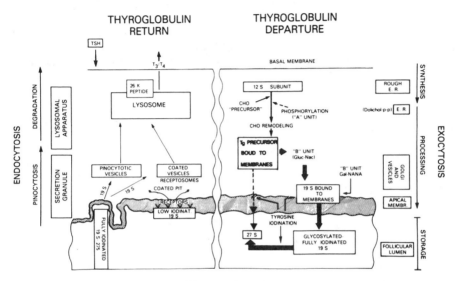

Figure 4

As this molecule progresses along the rough and smooth endoplasmic reticulum it undergoes its initial post-translational, glycosylation via dolichol phosphate intermediates. It is also in this area that phosphorilation of sugar and tyrosine residues can be anticipated to occur. It is at the end of this portion of the cell organelles, as it enters the Golgi apparatus, that remodeling of these initially synthesized oligosaccharide structures starts. Thyroglobulin, at this time, is present in the thyroid cell bound to membranes. The presence of "particulate" form of thyroglobulin bound to microsome fraction has been reported (20,21). However, preliminary data suggest that the thyroglobulin is bound to phospholipid and ganglioside moieties of the membrane. In the classical progression, further carbohydrate modification to form the B carbohydrate moiety by addition of GlcNAc and galactose stabilize the 19 S configuration (22), though the thyroglobulin remains membrane bound by virtue of the

GlcNac receptor recognition system. Vectorial movement through
the Golgi and exocytotic vesicles to the apical membrane with
continued completion of the B carbohydrate moiety; at the apical
membrane, sialation and iodination result in the release of the
glycosilated and iodinated final product into the follicular
lumen; which represents the storage compartment for the
iodoproteins.

With regards to endocytosis, it is presumed (10,12), that
poorly iodinated thyroglobulin molecules with partially
unglycosilated B chain moieties (14) are bound to the membrane
and unavailable to the TSH regulated pinocytotic process which
would thus tend to ungulf more fully iodinated and sialated
molecules. Receptosome internalization of these poorly
glycosilated intermediates in follicular thyroglobulin formation
could occur but is not believed at this time to be a
significant process. Pinocytotic vesicles merging with lisosome
and thyroglobulin degradation, perhaps with special intermediate
products of T_3 and T_4 rich 10, 20 and 26 K peptides, is the
presumed penultimate step to secretion of T_3 and T_4 from the
cell.

REFERENCES

1) Salvatore, G., Edelhoch, H. Chemistry and biosynthesis of thyroid iodoproteins. In Li CH (ed): Hormonal Proteins and peptide, Vol 1, pag. 201. New York, Academic Press, (1973).

2) Arima, T., and Spiro, R.G. Studies on the carbohydrate units of thyroglobulin. Structure of the mannose-N-acetyl glucosamine unit (unit A) of the human and calf thyroglobulin. J. Biol. Chem. 247, 1836-1848, (1972).

3) Arima, T. Spiro, M.J., and Spiro, R.G. Studies on the carbohydrate units of thyroglobulin. Evaluation of their microetherogeneity in the human and calf proteins. J. Biol. Chem. 247, 1825-1835, (1972).

4) Toyoshima, S., Fukuda, M., and Osawa, T. The presence of beta-mannosidic linkage in acidic glycopeptide from porcine thyroglobulin. Biochem. Biophys. Res. Commun. 51, 945-950, (1973).

5) Konfeld, R., and Kornfelf, S. Comparitive aspects of glycoprotein structures. Annu. Rev. Biochem. 45, 217-237, (1976).

6) Godelaine, D., Spiro, M.J., and Spiro, R.G. Processing of the carbohydrate units of thyroglobulin. J. Biol. Chem. 256, 10161-10168, (1981).

7) Nadler, N.J., Sarker, S.K., and Leblond, C.P. Origin of intracellular colloid droplets in the rat thyroid. Endocrinology, 71, 120-129, (1962).

8) Wollman, S.H. in Lysosomes in Biology and Medicine (Dingle, J.H., and Fell, H., eds) Vol 2 pp. 483-512, North Holland Publishing Co., Amsterdam. The Netherlans,(1969).

9) Salvatore, G., and Edelhoch, H. in Hormonal Proteins and Peptides (Li, C.H., ed) Vol 2, pp. 201-241, Academic Press, New York, (1973).

10) Cortese, F., Schneider, A.B., and Salvatore, G. Isopycnic centrifugation of thyroid iodoproteins: Selectivity of endocytosis. Eur. J. Biochem. 68, 121-129, (1976).

11) Gavaret, J.M., Deme, D., Nunez, J., and Salvatore, G. Sequential reactivity of tyrosil residues of thyroglobulin upon iodination catalyzed by thyroid peroxidase. J. Biol. Chem. 252, 3281-3285, (1977).

12) Wandenbroucke-van den Hove, M.F., De Vischer, M., and Salvatore, G. A new mechanism for the reabsorption of thyroid iodoproteins: Selective fluid pinocytosis Eur. J. Biochem. 122, 415-422, (1981).

13) Consiglio, E., Salvatore, G., Rall, J.E., and Kohn, D.L. Thyroglobulin interactions with thyroid plasma membranes. The existence of specific receptors and their potential role. J. Biol. Chem. 254, 5065-5076, (1979).

14) Consiglio, E., Shifrin, S., Yavin, Z., Ambesi-Impiombato, F.S., Rall, J., E., Salvatore, G., and Kohn, L.D. Thyroglobulin interactions with thyroid membranes: Relationship between receptor recognition of N-acetylglucosamine residues and the content of thyroglobulin preparations. J. Biol. Chem., 256, 10592-10599, (1981).

15) Shifrin, S. and Kohn L. D. Binding of thyroglobulin to bovine thyroid membranes: Role of specific aminoacids in receptor recognition. J. Biol. Chem. 256, 10600-10605, (1981).

16) Paulson, J.C., Hill, R.L., Tonabe, T., and Ashwell, G. Reactivation of asialo-rabbit liver binding protein by resialylation with B-DnGalactoside a 2-6 Sialyltransferase. J. Biol. Chem. 252, 8624-8628; (1977).

17) Monaco, F., Salvatore, G., and Robbins, J. The site of sialic acid incorporation into thyroglobulin in the thyroid gland. J. Biol. Chem. 250, 1595-1599, (1975).

18) Consiglio, E., Kohn, L. D., Salvatore, G., Shifrin, S., Cavallo, R., and Formisano, S. Mannose phosphate as a specific signal in the lysosomal degradation of thyroglobulin. In Advances in Neoplasia, Edited by M. Andreoli, F. Monaco and J. Robbins, pag. 61. Rome, Field Educational Italia, (1981).

19) Haslik, A. and Neufield, E.F., Biosynthesis of lysosomal enzymes in fibroblasts. Phosphorylation of mannose residues. J. Biol. Chem., 255, 4946-4950,(1980).

20) Vecchio, G., Claar, G.M. and Salvatore, G. Biosynthesis of thyroid iodoproteins in Vivo and in tissue slices. J. Biol. Chem. 247, 4908-4913, (1972).

21) Matsukawa,S. and Hosoya, T. Process of iodination of thyroglobulin and its maturation. J. Biochem. 85, 1009-1021 (1979).

22) Shifrin,S., Consiglio, E. and Kohn, L.D. Effect of the complex carbohydrate moiety on the structure of thyroglobulin. J. Biol. Chem., 258, 3780-3786, (1983).

GONADOTROPIN REGULATION OF SERTOLI CELL RESPONSE

M. Conti, M.V. Toscano, R. Geremia, P. Rossi,
L. Monaco, and M. Stefanini

Institute of Histology and General Embryology
University of Rome, via A. Scarpa 14, 00161 Rome, Italy

INTRODUCTION

Both gonadotropin and steroid hormones indirectly control the spermatogenic process by regulating the function of the Sertoli cell (Fritz, 1978). The peptide hormone FSH acts on Sertoli cell by binding to a specific receptor present on the plasma membrane of the cell (Means et al., 1978b). The hormone-receptor complex thus formed activates the adenylate cyclase system (Means et al., 1978a; Means et al., 1980) and produces an elevation of the intracellular cAMP (Dorrington et al., 1975). This increased cAMP concentration is, in turn, probably responsible for the several biological effects that have been observed after FSH stimulation of the Sertoli cell (Fritz, 1978; Means et al., 1978a,b).

The cAMP responsiveness to FSH is present only during a limited period of the lifespan of the Sertoli cell. Cells derived from immature testis respond readily to gonadotropin stimulation but, in spite of the presence of a full complement of gonadotropin receptors, adult Sertoli cells show a blunted response to FSH (Steinberger et al., 1978). This insensitivity to the gonadotropin appears to depend on FSH itself, since hypophysectomy of mature animals produces a return of the response to the "immature" levels (Means et al., 1978a). The physiological importance of these changes in Sertoli cell response is underscored by the observation that the sensitivity of the mature Sertoli cell undergoes a cycle. In fact, it has been suggested that adult Sertoli cells respond to

115

FSH to a variable degree which is dependent on the association with different stages of germ cell development (Parvinen et al., 1980). Thus, Sertoli cells associated with terminally differentiated spermatids do not respond to FSH, while cells associated with immature spermatids are highly responsive to the hormone. Similarly, the response of the Sertoli cells in vitro is dependent on the presence or absence of germ cells (Galdieri et al., 1981b). The mechanisms involved in the mentioned modulation of response are poorly defined. It is known that FSH treatment of Sertoli cell cultures produces also a refractory state (Verhoeven et al., 1980; O'Shaugnessy et al., 1980), but it is not known to what extent this refractoriness is similar to the modulation of response observed in the adult gonad. In the attempt to clarify the mechanisms that produce this phenomenon we have investigated the gonadotropin regulation of phosphodiesterase and the changes of phosphodiesterase activity that occur during testicular development. Here we report data on the phosphodiesterase stimulation by FSH, the changes of phosphodiesterase forms during Sertoli cell maturation, and finally the role played by this enzyme in the regulation of cell response.

EXPERIMENTAL DETAILS

    Male Wistar rats at different ages of development (between 10 and 60 days) were used in all experiments. In some experiments, fetuses were irradiated on the 19th day of pregnancy, according to the method of Means et al.(1976), to deplete germ cells population present in the seminiferous tubules. Seminiferous tubules were separated from the interstitium by mild collagenase digestion of testes freed of the capsula albuginea as previously reported (Conti et al., 1983a). Sertoli cell cultures were obtained as previously described (Conti et al., 1982) and freed of contaminating germ cells by short hypotonic treatment (Galdieri et al., 1981a). Phosphodiesterase activity was measured either after in vivo intraperitoneal injection of 50 µg oFSH (FSH-S14) in immature rats, or following various time intervals after treatment of Sertoli cell cultures with 0.5 µg/ml oFSH; in some instances 10^{-5} M isoproterenol or 0.5 mM dibutyryl cyclic AMP were used in place of gonadotropin. Variations in the activity of the hormone-sensitive adenylate cyclase were measured in vitro following various time intervals from addition of FSH or isoproterenol. Refractory state induced in vitro by hormone treatment was measured as cAMP production following a further hormonal stimulation (1 hr). The different forms of phosphodiesterase present in testis or testicular cell extracts were separated by DEAE-cellulose chromatography (Geremia et al., 1982). Phosphodiesterase and adenylate cyclase activities in the homogenates, and cAMP intracellular and extracellular levels were measured as described elsewhere (Conti et al., 1982; Conti et al., 1983b).

RESULTS AND DISCUSSION

FSH regulation of testicular phosphodiesterase

FSH stimulation of immature Sertoli cells in vitro causes a marked time- and dose-dependent increase in the phosphodiesterase activity measured in the cell homogenate. After a short term inhibition (30 min), the phosphodiesterase activity progressively increases between 1 hr and 12 hr and reaches a maximum following 24 hrs of treatment (Conti et al., 1981). Such stimulation can be reversed by removing FSH or the other stimulatory agent from the culture medium (Conti et al., 1982). The enzyme stimulation is probably mediated by the enhancement of cAMP intracellular levels. MIX does in fact potentiate FSH effect on phosphodiesterase, and the same induction can be detected after treatment with dibutyryl cAMP or other agents which increase cAMP levels, such as cholera toxin and beta-adrenergic agonists. Phosphodiesterase regulation requires also RNA and protein synthesis, since cycloheximide and actinomycin D block the hormonal induction of the phosphodiesterase activity (Conti et al., 1981; Verhoeven et al., 1981; Conti et al., 1982).

The FSH-dependent stimulation of phosphodiesterase is detected only when low cAMP concentration is used as substrate, while no changes are observed with cGMP (Conti et al., 1981). In agreement with these findings on total homogenate, chromatographic fractionation of cytosol from control and FSH-stimulated Sertoli cells demonstrated that the hormone induces a Ca^{2+} independent, cAMP specific high affinity isoenzyme, while the Ca^{2+} - calmodulin-dependent, cGMP specific form is not affected by the treatment (Conti et al., 1982).

A comparable effect of FSH on phosphodiesterase has been

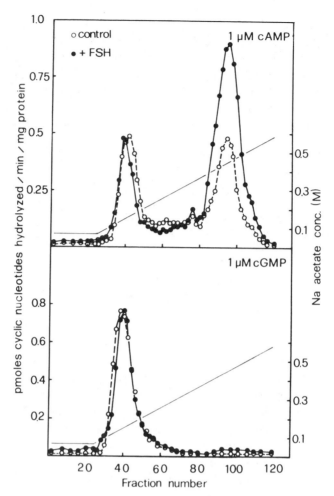

Fig. 1   Effect of FSH injection on the phosphodiesterase activity
          of SCE testis. Male rats irradiated in utero (Means et
          al., 1976) were injected intraperitoneally either with
          saline or with 50 μg oFSH on the 24th day from birth and
          were sacrificed 24 hrs later. Cytosol from seminiferous
          tubules was analyzed by DEAE-cellulose chromatography as
          described (Conti et al., 1983a). Phosphodiesterase activi-
          ty was assayed with either 1 μM cAMP (upper panel) or with
          1μM cGMP (lower panel) as substrates, in the presence of 1
          mM CaCl$_2$ and 100 ng/tube of purified calmodulin. The ac-
          tivity in the chromatograms from seminiferous tubule cyto-
          sol of control (○) and FSH-treated (●) animals was cor-
          rected for the amount of protein applied to the columns.

observed also in vivo. Intraperitoneal injection of FSH into 10 day-old immature male rats has been shown to produce a marked stimulation of cAMP hydrolytic activity of the seminiferous tubules (Conti et al., 1983a). FSH stimulation was present either when data were corrected per mg of protein or per testis, ruling out an aspecific hormonal effect on cell number or cell mass (Conti et al., 1983a). At the same time cGMP hydrolysis of the seminiferous tubule homogenate was not affected. The chromatographic analysis showed the selective augmentation of the same Ca^{2+}-calmodulin independent, cAMP specific phosphodiesterase induced by hormones in cultured Sertoli cells. It is probable that Sertoli cells are responsible for the induction observed in vivo, since they are target for FSH, and the enzymatic form stimulated by the hormone is present in the Sertoli cell extract but not in germ cells (Geremia et al, 1982). This interpretation is reinforced by the results of FSH treatment of immature animals with Sertoli cell enriched testis depleted of germ cells by X rays irradiation in utero. Also in this case, FSH injection produced a marked increase in the high affinity cAMP Ca^{2+}-calmodulin-independent enzyme (Fig.1).

Phosphodiesterase forms present in testicular cells and changes of activity during development

The data reported, as a whole, indicate that FSH administered both in vivo and in vitro regulates the phosphodiesterase activity in the Sertoli cell. It was also of interest to verify whether a correlation exists between the naturally occurring changes in FSH plasma levels and the phosphodiesterase activity of the Sertoli cell. It has been described, in fact, that plasma FSH levels increase between 20 and 30 day of age (Keteslegers et al., 1978) in temporal correlation with the development of Sertoli cell refractoriness to FSH (Means et al., 1980). A parallel variation of phosphodiesterase activity would suggest that cAMP catabolism is involved in the development of refractoriness in the maturing testis.

Fig. 2 shows the pattern of phosphodiesterase activity of cytosol from seminiferous tubules of 25 day old rats fractionated by DEAE-cellulose chromatography. The pattern of cAMP and cGMP hydrolytic activity, evaluated in the presence and absence of Ca^{2+} and calmodulin, revealed three main forms: a Ca^{2+}-calmodulin stimulated form eluting at 130 mM Na acetate which hydrolyzed both

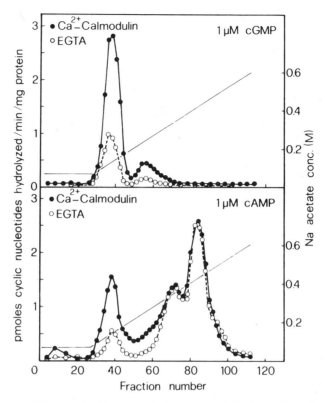

Fig. 2 Separation of the seminiferous tubule phosphodiesterase forms by ion-exchange chromatography. Seminiferous tubules prepared from 25 day old rats were homogenized and cytosol prepared and fractionated by DEAE-cellulose chromatography as described (Conti et al., 1983a). Every other fraction was analyzed either with 1 μM cGMP (upper panel) or with 1 μM cAMP (lower panel) as substrate, in the presence of EGTA (O) or excess Ca^{2+}-calmodulin (●).

cAMP and cGMP (peak I), and two Ca^{2+} independent forms which hydrolyzed specifically cAMP, eluting at 350 mM (peak II) and 430 mM salt (peak III) respectively. A minor Ca^{2+} stimulated activity hydrolysing both nucleotides eluted at 250 mM Na acetate. Interconversion experiments in which the enzyme activity was separated by ion-exchange chromatography in the presence or absence of Ca^{2+}, demonstrated that the Ca^{2+} stimulated activity eluting at 130 mM salt and that eluting at 250 mM salt represent the same enzyme which elutes at higher salt when bound to Ca^{2+} and calmodulin

(Geremia et al., 1982). The physico-chemical and kinetic parameters
which distinguish the three forms are reported in Tab.1.

Table 1. Properties of the different phosphodiesterase forms
present in the testicular cells.

	Peak I	Peak II	Peak III
Ionic strenght Na acetate (mM)	130	350	430
S_{w20}	6.8	6.5	3.9-5
App. K_m cAMP (μM)	20	4.5	1.7
App. K_m cGMP (μM)	3.4	ND	ND
Ca^{2+}-calmodulin stimulation	+	-	-

ND: not determined

The study of the distribution of the three isoenzymes in
different cell types of the testis was performed by using Sertoli
cell cultures (Conti et al., 1982), seminiferous tubules from
animals of various ages, and enriched germ cell populations obtained by albumin gradient fractionation (Geremia et al., 1982). The
results of these studies have indicated that peak I is present in
both somatic and germ cells, peak II is specific of germ cells, and
peak III is present only in Sertoli cells (Geremia et al., 1982,
1983; Conti et al., 1982). The variations of phosphodiesterase
activity in Sertoli cells were studied in homogenates both from
interstitium-free seminiferous tubules and from Sertoli cells
cultures from animals of different ages. Both cAMP specific forms,
and, to a lesser extent, the cGMP specific form, increased with the
age of the animal. While the increase of the germ cell specific
form is the result of the increase of germ cell number in the
developing testis, the augmentation of the Sertoli cell specific
form is probably due to a modification of the specific activity in

this cell type (Geremia et al., 1982, 1983), since Sertoli cell number, in the rat, remains unchanged after approximately 15 days of age. This result is confirmed by the studies performed on Sertoli cell culture from animals at different ages. While the specific activity of the cGMP specific form remains unchanged with testicular development, the activity of the cAMP specific phosphodiesterase is increased up to 10 fold from 10 to 30 days of age (paper in preparation). The described increase of PDE correlates well with the increase of plasma FSH levels that occur at 20 to 30 days of age in the rat (Keteslegers et al., 1978), thus suggesting that one of the physiological effects of FSH on the Sertoli cell is the induction of phosphodiesterase activity. These changes in phosphodiesterase activity might therefore be relevant for the development of the refractory state of the maturing Sertoli cell (Means et al., 1978a,b).

Involvement of phosphodiesterase in the "in vitro" refractory state of the Sertoli cell

Previous studies have indicated that receptor down-regulation (O'Shaugnessy, 1980) and adenylate cyclase desensitization (van Sickle et al., 1981) are involved in the reduction of Sertoli cell response to gonadotropin stimulation. Once established that FSH regulates phosphodiesterase activity, attempts have been made to assess whether this enzyme is involved in the onset of refractoriness which follows hormonal stimulation. Evidence obtained in vitro on the kinetics of refractoriness in cultured Sertoli cells suggests that this might be the case (Conti et al., 1983b): 1) an FSH-dependent regulation of cAMP catabolism is implied by the high affinity for cAMP of the enzyme stimulated by the hormone; 2) the time course of the stimulation of phosphodiesterase by FSH in cultured Sertoli cells corresponds to that of the onset of the refractory state; 3) while all the agents that induce a refractory state stimulate the phosphodiesterase activity, not all of them can elicit adenylate cyclase desensitization; 4) phosphodiesterase inhibitors, such as MIX, can partially restore the hormonal stimulation of FSH-dependent cAMP accumulation. Moreover, partial prevention of the onset of FSH-dependent refractoriness can be obtained by adding cycloheximide to culture dishes during hormonal pretreatment. Since protein synthesis inhibitors can prevent FSH-induction of phosphodiesterase activity, while they have no influence on adenylate cyclase desensitization, we can conclude that the cycloheximide restores the   response to gonadotropin by blocking the phosphodiesterase induction. On the other hand, these studies suggest that both phosphodiesterase induction and adenylate cyclase desensitization are needed for the onset of refractoriness, since MIX and cycloheximide cannot restore completely FSH-dependent cAMP accumulation. When very low doses of gonadotropin are used (nanogram range), refractoriness of the cell is associated only with stimulation of phosphodiesterase and not with desensitization of adenylate cyclase (Conti et al., 1983b).

Sertoli cells respond not only to FSH but also to beta-adrenergic agents, such as isoproterenol, via a distinct receptor (Verhoeven et al., 1979). FSH-induced refractory state also impairs isoproterenol-dependent cAMP accumulation in the Sertoli cells (Verhoeven et al., 1981; Conti et al., 1983b). This heterologous refractoriness is not accompanied by desensitization of beta-adrenergic-dependent adenylate cyclase. Also FSH response of the Sertoli cell in terms of cAMP accumulation is inhibited by isoproterenol pretreatment, whereas FSH-sensitive adenylate cyclase still responds to gonadotropin stimulus (Conti et al., 1983b). In both cases phosphodiesterase activity is increased by hormonal treatments, thus suggesting that the induction of this enzyme is the only factor responsible for the onset of heterologous refractoriness.

ACKNOWLEDGEMENTS

The authors are thankful to R. Pezzotti and to L. Petrelli for the skillful technical assistance. The work reported was supported by grants of the Ministero della Pubblica Istruzione and by CNR targeted project "Medicina Preventiva e Riabilitativa". Dr. P. Rossi is a recipient of a postdoctoral fellowship of the "Valerio Monesi" Foundation.

REFERENCES

Conti, M., Geremia, R., Adamo, S. and Stefanini M., 1981, Regulation of Sertoli cell cyclic adenosine 3':5' monophosphate phosphodiesterase activity by follicle stimulating hormone and dibutyryl cyclic AMP, Biochem. Biophys. Res. Commun., 98:1044.

Conti, M., Toscano, M. V., Petrelli, L., Geremia, R. and Stefanini, M., 1982, Regulation by follicle stimulating hormone and dibutyryl adenosine 3':5' monophosphate of a phosphodiesterase isoenzyme of the Sertoli cell, Endocrinology, 110:1189.

Conti, M., Toscano, M. V., Geremia, R. and Stefanini, M., 1983a, Follicle-stimulating hormone regulates in vivo testicular phosphodiesterase, Mol. Cell. Endocrinol., 29:79.

Conti, M., Toscano, M. V., Petrelli, L., Geremia, R. and Stefanini, M., 1983b, Involvement of phosphodiesterase in the refractoriness of the Sertoli cell, Endocrinology, 113:1635.

Dorrington, J. M., Roller, N. F. and Fritz I. B., 1975, Effects of follicle stimulating hormone on cultures of Sertoli cell preparations, Mol. Cell. Endocrinol., 3:57.

Fritz, I. B., 1978, Sites of action of androgens and follicle stimulating hormone on cells of the seminiferous tubule, in: "Biochemical Actions of Hormones", G. Litwoch, ed., vol. V, pp. 249-281, Academic Press, New York, N.Y.

Galdieri, M., Ziparo E., Palombi, F., Russo, M. A. and Stefanini, M., 1981a, Pure Sertoli cell cultures: a new model for the study of somatic-germ cell interaction, J. Androl., 5:249.

Galdieri, M., Zani, B. and Stefanini, M., 1981b, Effect of the association with germ cells on the secretory activity of Sertoli cells in vitro cultures, in: "Oligozoospermia: Recent Progress in Andrology", G. Frajese et al., eds., pp. 95-103, Raven Press, New York, N. Y.

Geremia, R., Rossi, P., Pezzotti, R. and Conti, M., 1982, Cyclic nucleotide phosphodiesterase in developing rat testis. Identification of somatic and germ cell forms, Mol. Cell. Endocrinol., 28:37.

Geremia, R., Rossi, P., Conti, M. and Stefanini, M., 1983, Regulation of testicular phosphodiesterases during development and hormone treatments, in: "Recent Advances in Male Reproduction. Molecular bases and Clinical Implications", R. D'Agata, M. B. Lipsett and H. J. Van der Molen, eds., pp. 121-128, Raven Press, New York, N.Y.

Keteslegers, J. M., Hetzel, W. D., Sherms, R. J. and Catt, K. J., 1978, Developmental changes in testicular gonadotropin receptors, plasma gonadotropins and plasma testosterone in the rat, Endocrinology, 103:212.

Means, A. R., Fakunding, J. L., Huckins, C., Tindall, D. J. and Vitale, R., 1976, Follicle-stimulating-hormone, the Sertoli cell and Spermatogenesis, Recent Prog. Hormone Res., 32:477.

Means, A. R., Dedman, J. R., Tindall, D. S. and Welsh, M. J.,

1978a, Hormonal regulation of Sertoli cells, <u>Int. J. Androl.</u>, suppl. 2:403.

Means, A. R., Dedman, J. R., Fakunding, J. L. and Tindall, D. J., 1978b, Mechanism of action of FSH in the male rat, <u>in</u>: "Receptor and Hormone Action", L. Birnbaumer and B. W. O'Malley, eds., vol. III, pp. 363-393, Academic Press, N.Y.

O'Shaughnessy, P. J., 1980, FSH receptor autoregulation and cyclic AMP production in the immature rat testis, <u>Biol Reprod.</u>, 23:810.

Parvinen, M., Marana, R., Robertson, D. M., Hansson, V. and Ritzen, E.M., 1980, Functional cycle of rat Sertoli cells: differential binding and action of follicle-stimulating hormone at various stages of the spermatogenic cycle, <u>in</u>: "Testicular Development, Structure and Function", A. Steinberger and E. Steinberger eds., pp. 425-432, Raven Press , New York.

Steinberger, A., Hintz, M. and Heindel, J. J., 1978, Changes in cyclic AMP responses to FSH in isolate rat Sertoli cells during sexual maturation, <u>Biol. Reprod.</u>, 19:566.

Van Sickle, M., Oberwetter, J. M., Birnbaumer, L. and Means, A. R., 1981, Developmental changes in the hormonal regulation of rat testis Sertoli cell adenylyl cyclase, <u>Endocrinology</u>, 109:1270.

Verhoeven, G., Dierickx, P. and de Moor, P., 1979, Stimulation effect of neurotransmitters on the aromatization of testosterone by Sertoli cell enriched cultures, <u>Mol. Cell. Endocrinol.</u>, 13:241.

Verhoeven, G., Cailleau, J. and de Moor, P., 1980, Desensitization of cultured rat Sertoli cells by follicle stimulating hormone and by L-isoproterenol, <u>Mol. Cell. Endocrinol.</u>, 20:113.

Verhoeven, G., Cailleau, J. and de Moor, P., 1981, Hormonal control of phosphodiesterase activity in cultured rat Sertoli cells, <u>Mol. Cell. Endocrinol.</u>, 24:41.

NEW CLASS OF OPIOID PEPTIDES: DERMORPHINS

P. Melchiorri, G. Improta, L. Negri and M. Broccardo

Institute of Pharmacology, University "La Sapienza"
Rome, Italy

INTRODUCTION

In recent years, a number of biologically active peptides have been isolated first form the skin of amphibian species and then traced in gastrointestinal endocrine cells and in central and peripheral neurons of mammals (Erspamer et al., 1981). In fact, the discovery of a new skin peptide appears to be prerequisite to the identification of analogous peptides in mammalian tissues. Striking examples are caerulein, tachykinins, bombesin, xenopsin, sauvagine, and recently dermorphin.

Studies on skin peptides have not only heralded the discovery of new brain and gut peptides, but have substantially aided in elucidating their structure (tachykinins and substance P), in determining the essential amino acids in the peptide molecule (caerulein and colecystokynin) and in providing important models for studying structure-activity relationships. Moreover, the early availability of synthetic amphibian peptides, sometimes preceding the corresponding mammalian peptides by several years, has permitted the performance of extensive pharmacological studies, which have then proved to be valid for all members of a given family, independently of their origin.

The simultaneous occurrence of several peptide families in the brain, GI tract and skin of frogs, led to formulate the concept of the existence of a brain-gut-skin peptide triangle (Erspamer et al., 1983). It will be interesting to ascertain whether the validity of this hypothesis is limited to amphibians or whether mammalian brain, gut and skin may also contain the same active peptides, at least during embryonic development or in the case of neoplastic growth.

Until now, the isolated endogenous opioid peptides can be grouped into three classes: B-endorphin and certain related compounds; the enkephalins which are the smallest opioid peptides; and dynorphin and α-neo-dynorphin. Each group has its distinct precursors and may have distinct or partially distinct distribution in the brain. They also seem to interact with different receptors in the brain. Enkephalins bind mainly to δ-receptors, which are distinct from the μ-receptors binding morphine, while β-endorphin seems to interact with both types. The binding of dynorphin in brain has not been well studied, but in vitro studies in the guinea-pig ileum indicate it binds to another receptor, the K-receptor.

Dermorphins are a new class of opioid peptides isolated from the skin of several frogs of the genus Phyllomedusa (Montecucchi et al. 1981 a,b). Like other peptides found in amphibians, they have profound effects on many mammalian systems. Furthermore, dermorphin-like immunoreactivity has been already traced in mammalian brain (Negri et al., 1981).

The purpose of this paper is twofold: (i) to summarize pharmacological data available on this class of opioid peptides; and (ii) to provide evidence for the presence of dermorphin-like peptides in the nervous system of higher vertebrates.

CHEMISTRY

Dermorphins, isolated from methanol extracts of the skin of the South American frogs Phyllomedusa sauvagei, Phyllomedusa rhodei and Phyllomedusa burmeisteri, are two heptapeptides whose primary structure present the unique feature of having a D-amino acid residue in position 2 (D- Alanine), which is critical for biological activity

Tyr–ala–Phe–Gly–Tyr–Pro–Ser–NH$_2$ Dermorphin

Tyr–ala–Phe–Gly–Tyr–Hyp–Ser–NH$_2$ Hyp^6 – dermorphin

On the basis of the available evidence we are convinced that the
sequences of dermorphins actually exist in the living amphibian
skin and are not artifacts resulting from total inversion of L–Ala^2
dermorphin during extraction and purification procedures.

OCCURRENCE AND DISTRIBUTION

Dermorphins appear to be distributed widely in the animal
kingdom, from marine mollusca to terrestrial mammals, but in
extended range of concentrations, from few ng/g to several mcg/g of
tissue. (Negri et al, 1981, see Table 1).

Table 1. Dermorphin-like peptide content of different animal species.

Octopus macropus	(optical ganglia)	0.80 ng/g
Eledone moscata	(optical ganglia)	0.70 ng/g
Dosidicus gigas	(mantle nerve)	0.35 ng/g
Phyllomedusa sauvagei	(skin)	45.ug/g
Phyllomedusa rhodei	(skin)	100 ug/g
Phyllomedusa burmeisteri	(skin)	80 ug/g
Physalemus bigillonigerus	(skin)	15 ug/g
Engistomps pustulosus	(skin)	550 ug/g
Bufo viridis	(skin)	140 ng/g
Pleuroderma cinerea	(skin)	50 ng/g
Rana esculenta	(skin)	12 ng/g
Rana esculenta	(brain)	8 ng/g
Rana esculenta	(eye)	35 ng/g
Rat brain		1–5 ng/g
Pig brain		0.8–2 ng/g

The presence of dermorphin-like peptide(s) in animal tissues
different from amphibian skin was assessed by radioimmunoassay and
immunohistochemistry, but no effort succeeded in isolating and
sequencing such small amounts of peptides. Their structure,
therefore, remains still unknown.

Quite recently Buffa et al. (1982) studied the distribution
of dermorphin-like peptide(s) in rat brain by immunohistochemistry.
Intense staining of immunoreactive dermorphin was found in nerve
cell bodies and fibres of the paraventricular nucleus and arcuate
nucleus of the hypothalamus, around the organum vasculosum of the
lamina terminalis, in the subfornical organ, the supraoptic
nucleus, the nucleus ambiguus, the nucleus tractus solitarium, the
hypoglossal nerve nucleus, and in the nucleus olivaris inferior.
Dermorphin neurons were of short or intermediate length and
immunostaining was prevented by absorption of diluted immunoserum
with $(D-Ala^2)$-dermorphin, but not with $(L-Ala^2)$-dermorphin or other
homologous or peptides. The distribution of dermorphin-like
immunoreactivity within rat brain appears to be quite different
from that of other opioid-related molecules, as enkephalins,
endorphins and ACTH. On the other hand, the concentration of
dermorphin immunoreactivity in the hypothalamus, coupled with its
lack in median eminence, suggests some role of dermorphine in local
modulation of hypothalamic structures, with special reference to
the arcuate and OVLT regions. It seems pertinent to outline that
high concentrations of LHRH have been found in the arcuate nucleus
and suprachiasmatic area inclunding OVLT. In fact, a strong
inhibitory activity of dermorphin on LH secretion has been recently
observed (Gullner et al., 1983). The presence of dermorphin
immunoreactivity in the nucleus tractus solitarius and in a
dorso-medial area of the medulla, just beneath the floor of the IV
ventricle, may provide again the rationale to suggest a modulating
role of the peptide on these structures. In fact, intracerebroventricular injection of dermorphin in rats produces a strong
inhibition of gastric acid secretion evoked by gastric distension
and a long lasting hypotension (Improta et al. 1982), while the
solitary nuclei of the brain stem receives sensory input from the
vagus nerve, contains opiate receptors (Kuhar, 1978) and has a well
established role in regulating central mediated gastric secretion
and blood pressure.

PHARMACOKINETICS

Following i.v. injection in rats, dermorphin was rapidly removed from blood and destroyed in the liver and kidney; its half-life averaged 1.3 min. Approximately 80% of the peptide that reached the liver was cleared in a single passage, and an aliquot of unmodified peptide (15%), together with its inactive fragments eliminated in the bile. Only 0,5% of injected dermorphin was recovered in rat urine. Less than 0.01% of injected dermorphin passed rat blood-brain barrier.

The proteolytic enzyme system involved in the cleavage of dermorphin by the liver and kidney produced as the main degradation product the N-terminal tripeptide fragment Tyr-D-Ala-Phe, which is devoid of any biological activity (Negri et al., in press).

PHARMACODYNAMICS

The present status of pharmacological research on dermorphins is summarised below.

Opiate Receptor Binding

Dermorphin and (Hyp^6)-dermorphin caused a 50% inhibition of (^{3}H) Leu-enkefalin (7 nM) binding to opioid receptors in membranes of mouse neuroblastoma and rat glioma hybrid cells at concentrations of 0.1 and 0.3 μM, respectively. The L-Ala^2 analogue of dermorphin was more than 100 times less potent, while Leu-enkefalin was 10 times as potent as dermorphin. In addition, like other opioid peptides, dermorphin and (Hyp^6)-dermorphin inhibited the prostaglandin E$_1$-induced increase in cyclic AMP levels. IC$_{50}$ values were 0.2 and 0.4 μM, respectively. Leu-enkephalin was 2000 times as potent as dermorphin, but (L-Ala^2) dermorphin was at least 1000 times less potent. Naloxone blocked the inhibitory effect of dermorphin. The greater potency of Leu-enkephalin compared to the dermorphins is likely due to the fact that hybrid cells are carrying δ-type opiate receptors, upon which dermorphins are much less potent than on μ-type receptors. In fact, in a specific binding assay for μ-opiate receptors (rat brain membranes-^{3}H-sufentanil) dermorphin had an affinity (IC_{50}) of 0.6 nM, whereas in a specific binding assay for δ-opiate receptors (cultured NG 108-5 cells ^{3}H-DADL) it had an affinity of only 160.0 nM. Doble et al. (1982), in turn, observed that dermorphin had no effect on the opiate receptors of the clam rectum, resembling the vertebrate δ-receptors.

Central Nervous System

   Dermorphin produced considerable analgesia in mice by i.v. injection (ED_{50} 1.02 μmoles/kg) and in rats by i.c.v. injection (ED_{50}=50 pmoles/Kg). Duration of analgesia was considerably longer than that reported in the literature for B-endorphin. Pretreatment of the animals with naloxone (3 μmoles/kg, s.c.) completely abolished the analgesic effect of the peptide. Table 2 shows that no natural compound can compete with dermorphin in its central antinociceptive action. β-Endorphin, the most potent agent, showed barely 4-6% of the activity of dermorphin, morphine 0.05-0.13%, the two enkephalins, (1,13)-dynorphin and B-casomorphin-7 less than 0.1%. Intracerebroventricular doses of dermorphin exceeding 500 pmoles/kg produced in the rat catalepsy and intense rigidity (Tab. 2 and Fig. 1 and 2).

   To localize the central sites in the rat related to some dermorphin pharmacological effects, namely analgesia and catalepsy, it was microinjected bilaterally into different brain areas and the effects were assessed at 30 min intervals for 2 h. Analgesia was produced at the lowest dose level (ED_{50} 0.6-1 pmole/rat) when the peptide was injected into the anterior hypothalamus and the periaqueductal gray-4th ventricular spaces. Catalepsy, on the contrary, was most evident when dermorphin was injected into the nucleus accumbens of the preoptic area (Melchiorri et al., 1983).

   Injected i.c.v. in rabbits, dermorphin produced an initial desynchronization of EEG, followed by a synchronous EEG pattern associated with muscular rigidity, exophthalmus, mydriasis and analgesia. At doses of 3-10 nmoles the rapid appearance of the synchronous pattern, with high voltage slow waves in the cortical leads, was accompanied by complete disruption of "theta" waves in the hippocampal lead and by attenuation or blockade of the EEG arousal response. Behaviourally, the animals showed marked muscular rigidity, catalepsy and loss of the righting reflex. Spike wave

complexes and isolated spikes appeared only at doses of 25-35
nmoles (Aloisi et al., 1982).

I.c.v. injections of dermorphin in rats in doses ranging from
6 to 1200 pmoles/kg caused more or less sustained grand mal EEG
seizures associated with twitches and slight tremors at the lower
doses and tonic-clonic convulsions at the higher doses. The
convulsive patterns were followed by high voltage slow waves,
interrupted by periods of low voltage desynchronous EEG. According
to Aloisi et al (1982), dermorphin was 2500 and 10000 times as
potent as morphine and the enkephalins, respectively, in inducing
epileptic fits in rats.
Dermorphin given by i.c.v. injection at doses ranging from 12.5 to
125 pmoles/mouse elicited a typical Straub-tail reaction,
accompanied by restlessness and increased motor activity. Doses
larger than 125 pmoles produced only catalepsy. Naloxone completely
antagonized these effects (Aloisi et al., 1982).

These results were confirmed and extended by Puglisi-Allegra
et al. (1982), measuring the locomotor activity with toggle floor
boxes. It was shown that dermorphin injected into the lateral
ventricles (0.03-1.1 nmoles/mouse) induced a dose related increase
in the running activity which lasted more than 2 h and was similar
to the "running fit" induced by morphine, which was at least 50
times less potent than dermorphin. However, given by the i.v.
route, dermorphin elicited in mice an opposite effect, i.e.,
depression of motor activity.

Dermorphin instilled into the 3rd ventricle of chicks produced
a dose-dependent behavioural and electrocortical sedation or sleep
and a decrease in body temperature. The threshold dose of 1.25
pmoles/chick produced only slight sedation lasting about 10 min;
larger doses (up to 125 pmoles/chick) caused profound behavioural
and electrocortical soporific effects ("squatting"), lasting up to
4 h. Power spectrum analysis of EEG under dermorphin revealed a
sustained increase in total voltage power as well as a shift
towards the lower frequency band (0-3 Hz). Naloxone reversed both
behavioural and electrocortical effects. On a molar basis
dermorphin was 24 times as potent as B-endorphin in producing sleep
of similar duration and 27 times as potent in causing a similar
increase in total voltage power and a shift towards the low
frequency band. Since dermorphin was found to be an extraordinarily
potent peptide in inducing sleep and since it acts at much lower

Table 2. Relative potency, on a molar basis, of dermorphin, other natural opioid peptides and morphine, in 4 in vivo and in vitro test preparations (dermorphin = 1000).

	Guinea pig ileum	Mouse vas deferens	Rat hot plate test[a]	Rat tail flick test[a]
Dermorphin	100	100	100	100
Hyp⁶-dermorphin	80 – 90	85 – 90	90 – 110	70 – 80
Met-enkephalin	1.5 – 2	100 – 120	<0.01	<0.01
Leu-enkephalin	0.35	100 – 130	<0.01	<0.01
β-Endorphin	5 – 7	6	–	4 – 6
(1 – 13)-Dynorphin	300 – 500	40 – 70	–	0.1
β-Casomorphin-7	0.05	0.15 – 0.4	<0.1	–
Morphine	2.5	2.5	0.05	0.13

[a] I.c.v. injection; – , not determined.

Fig. 1. Time-response curves for analgesic activity in hot plate test of different doses of dermorphin, given by i.v. injection to groups of 10 mice each. Vertical bars represent S.E.M. Reproduced from Broccardo et al. (1981) Br. J. Pharmacol. 73, 625-631.

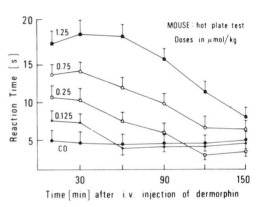

Fig. 2. Time-response curves for analgesic activity of a single dose of dermorphin administered i.c.v. in groups of 10 animals each. Responses obtained in both the hot plate test and the tail flick test are shown. Vertical bars represent S.E.M. Reproduced from Broccardo et al. (1981) Br. J. Pharmacol. 73, 625-631.

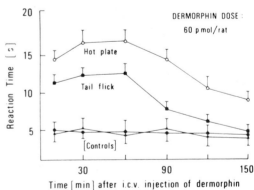

than other real or putative transmitters, it is suggested that, at least in birds, it may play a role as a mediator of sleep (Nisticò et al., 1982). (Fig. 3).

Dermorphin i.c.v. in rats at at doses of 1-5 nmoles/Kg produced a 35-40% increase in the dehydroxyphenylacetic acid concentration in corpus striatum, whereas the DA concentration was unchanged. Naloxone antagonized this effect (Gessa, personal communication).

Tolerance and physical dependence

By using osmotic micropumps, Broccardo et al. (1983) infused dermorphin into lateral ventricle of rat brain for a tow-week period and studied the development of tolerance to the analgesic and cataleptic action of the peptide as well as physical dependence. Tolerance to catalepsy and rigidity was established within 4 h, while analgesia dissapeared after 3-day infusion. Development of physical dependence was complete at the 3rd day of treatment; after naloxone injection the most frequent and characteristic withdrawal signs were: repeated attempts of the animal to escape from its cage, exploring, teeth chattering, the frequency of grooming and shaking increased until the 10th day.

Pheripheral organs

In vitro preparations. Dermorphin displayed a potent and long-lasting depressive activity on electrically stimulated contractions of guinea pig ileum and mouse vas deferens preparations (Broccardo et al., 1981). Naloxone was a powerful antagonist. In the mouse vas deferens preparations, naloxone was considerably more potent in blocking dermorphin than in blocking the enkephalins or dynorphin. The relative potencies of dermorphin, other natural opioid peptides of the endorphin-enkephalin and exophin series, and of morphine are shown in Table 2.

Pituitary gland. Dermorphin produced in the rat a marked release of prolactin, as shown by the rise in serum prolactin levels (Motta et al., 1983). The threshold dose by s.c. administration was 0.2-0.3

µmoles/kg, and the effect was dose-dependent. Pretreatment with
naloxone (2 mg/Kg i.p.) completely prevented the rise in serum

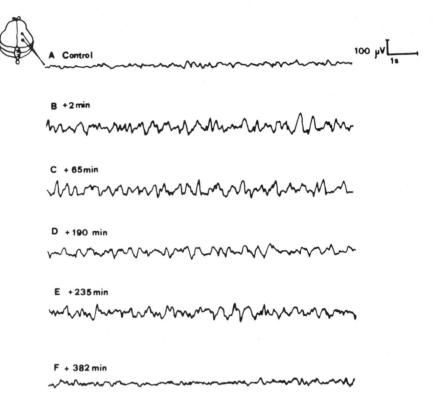

Fig. 3. Electrocortical effects of an intraventricular
administration of 0.22 nmole dermorphin in the chick. A, control
ECG activity; B, C, D, E, F, ECG activity at various times after
dermorphin injection. The peptide produced an increase in the
amplitude potentials and a decrease in the ECG frequency. Courtesy
of Dr. G. Nisticò.

Prl induced by 1 µmole/kg of dermorphin. In normal male rats
injection of 0.25 µg/kg of dermorphin was not able to induce any
significant change in serum Prl levels 10 min after injection.
Serum Prl levels showed a significant enhancement 30 min after the
administration of this dose of dermorphin, and returned to control
values at the 60 min interval. Conversely, 1 µg/kg of dermorphin
significantly elevated Prl concentrations 10 min after injection.

Serum Prl levels showed a significant enhancement 30 min after the administration of this dose of dermorphin, and returned to control values at the 60 min interval. Conversely, 1 µg/kg of dermorphin significantly elevated Prl concentrations 10 min after injection. In the isolated and dispersed rat pituitary cell preparation, dermorphin, added to the medium at a concentration of 5×10^{-7} M, did not induce any alteration in Prl output. The peptide was also unable to counteract the inhibiting effect of a continuous perfusion with DA on Prl release. These data provide evidence that dermorphin influences Prl release through an action which takes place in the central nervous system and possibly in the hypothalamus. (Motta et al., 1983).

Dermorphin, given by the i.c.v. route, produced a dose-dependent inhibition of LH release in long-term castrated adult male rats, as shown by a significant decrease in serum immunoreactive LH levels.

Dermorphin did not affect FSH release, but increased TSH release (Gullner et al., 1983).

Stomach. Dermorphin injected i.c.v. into pylorus-ligated rats was an extremely potent inhibitor of gastric secretion induced by water distension of the stomach (Fig. 4). At dose levels ten times higher, dermorphin injected i.c.v. in rats produced a dose-related delay in gastric emptying which was antagonized by naloxone (Fig.4). When administered by microinjection directly into the brain tissue, dermorphin displayed complex effects depending on the site of injection and dose. High doses of the peptide (20-40 pmoles/kg) injected into the lateral hypothalamus increased basal gastric volume and acid output, but had no effect on acid secretion stimulated by gastric distension or 2-deoxy-D-glucose injection. Low doses (1-6 pmoles/kg) injected into nucleus solitarium or 4th ventricle blocked distension-stimulated gastric secretion and reduced 2-deoxy-D-glucose-induced acid secretion (Improta et al., 1982).

Dermorphin given by i.v. infusion, at doses of 1.2, 2.5 and 3.7 nmoles/kg/h, produced a dose dependent reduction of gastric output elicited by 2-deoxy-D-glucose. Maximum reduction occurred during the first hour and was approximately 50%. In man, dermorphin administered s.c. at a dose level of 2 mg reduced gastric acid secretion both in normal volunteers and in subjects submitted to truncal vagotomy (Lezoche et al., 1983). Gastric

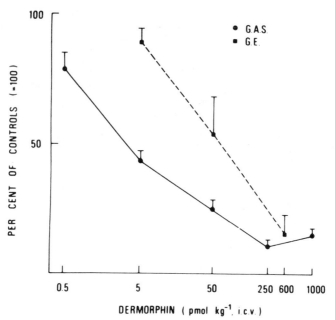

Fig. 4. Gastric acid secretion (G.A.S.) in conscious, pylorus-ligated rats, given a water load, and gastric emptying (G.E.) in conscious rats. The inhibitory effect of increasing doses of dermorphin, administered by i.c.v. route, is shown, as percentage of controls. Group of 5 rats were used. Vertical bars represent S.E.M.

emptying in rats was also reduced by i.c.v. injections of dermorphin (Broccardo et al., 1982), (Fig.4).

SYNTHETIC DERMORPHIN ANALOGUES

Approximately 100 dermorphin homologues and analogues have been synthesized (De Castiglione et al., 1981). The relative potency of some of them is shown in Table 3. The L-Ala^2-dermorphin is virtually ineffective on all preparations tested. Substitution of D-Ala^2 with D-Met^2 causes a sharp reduction of biological activity. The minimum sequence requirement for biological activity is represented by N-terminal tetrapeptide. Benzylation of Tyr^5 and/or Ser^7 in dermorphin produces an increased activity on mouse vas deferens, but a sharp decrease of the effects on CNS and on guinea pig ileum. Deaminated peptides, while retaining good

Table 3. Relative potency, on a molar basis, of synthetic dermorphin analogues in 4 in vitro and in vivo test preparations (dermorphin = 100).

	Guinea pig ileum	Mouse vas deferens	Rat hot plate test[a]	Rat tail flick test[a]
Dermorphin	100	100	100	100
Dermorphin-(1 – 6)	44 – 52	35 – 40	25	14
Dermorphin-(1 – 5)	40 – 45	54 – 62	27	12
Dermorphin-(1 – 4)	2.5 – 3.5	3 – 3.5	22	12
Dermorphin-(1 – 3)	<0.03	<0.05	<0.2	<0.1
Dermorphin-(5 – 7)	<0.03	0.1	<0.02	–
[L-Ala^2]Dermorphin	<0.03	0.06	<0.2	
[D-Met^2]Dermorphin	2.5 – 3	6 – 7	<0.02	<0.02
[Gly^3]Dermorphin	<0.03	0.1	<0.02	–
[Gly^3,Phe^4]Dermorphin	8 – 9	15 – 20	1	22 –
[Phe^4]Dermorphin	25 – 30	25 – 30	10	–
[Sar^4]Dermorphin	110 – 140	110 – 120	100	70
[Pro^4]Dermorphin	1	1.5	3	1
[Gly^5]Dermorphin	0.4 – 0.5	0.5 – 1	0.5	0.1
[Phe^5]Dermorphin	40 – 50	50 – 55	70	10 – 20
[Trp^5]Dermorphin	50 – 60	50 – 60	30	3.5
[Phe^5,Hyp^6]Dermorphin	70 – 80	50	70	5
[Val^6]Dermorphin	16 – 20	17 – 20	5	–
[Gly^6]Dermorphin	30	12 – 15	10	–
[Δ-D-Pro^6]Dermorphin	13 – 15	14 – 16	–	–
[Gly^7]Dermorphin	70 – 80	65 – 75	20	8
[Abu^7]Dermorphin	70 – 80	65 – 75	–	–
Deamidated dermorphin	23 – 30	60 – 70	0.5	1
[Tyr(Bzl)^5]Dermorphin	4	40 – 50	2	0.5
[Ser(Bzl)^7]Dermorphin	70 – 90	270 – 300	20	8
[Tyr(Bzl)^5,Ser(Bzl)^7]Dermorphin	3.5 – 4	250 – 500	0.5	1
[Tyr(Bzl)^5,Hyp^6,Ser(Bzl)^7]Dermorphin	12 – 18	300 – 1500	10	–
[Boc]Dermorphin	0.3 – 0.4	0.15	<0.02	<0.02
[Des-Pro^6]Dermorphin	30 – 40	35 – 50	8	1
[Des-Tyr^5]Dermorphin	0.1	0.1 – 0.2	2.5	1
[Des-Gly^4]Dermorphin	0.4 – 0.6	1	–	–
[Leu^5]Dermorphin-(1 – 5)	15 – 17	150 – 190	<0.5	<0.1
[Met^5]Dermorphin-(1 – 5)	14 – 16	33 – 36	–	–
[Pro^5]Dermorphin-(1 – 5)	0.1 – 0.2	0.5 – 1	–	–
[D-Tyr^5]Dermorphin-(1 – 5)	17 – 19	190 – 230	5	1.5
[Tyr-NH-NH$_2$^{5}]Dermorphin-(1 – 5)	24 – 32	30 – 35	20 – 30	–
[Tyr-NH-NH-Z^5]Dermorphin-(1 – 5)	50 - 60	70 – 75	20 – 40	12

[a] I.c.v. injection.

– , not determined; Abu, L-2-aminobutyric acid; Boc, t-butyloxycarbonyl; -NH-NH$_2$, hydrazide; -NH-NH-Z, carbobenzoxyhydrazide.

Reproduced from De Castiglione et al. (1982) and Melchiorri et al. (1982).

activity on guinea pig ileum and mouse vas deferens preparation, are virtually inactive on CNS.

In enkephalins, the phenolic ring and amino terminus of Tyr^1 and the Phe^4 (or other similar hydrophobic) residue are considered to fulfill a three-site requirement for biological activity (Gorin et al., 1980), while Gly^3 appears to provide a degree of flexibility to the molecule (Beddell et al., 1977). In dermorphins, the phenolic hydroxyl group of the first amino acid residue does not seem to be an absolute requirement, at least for peripheral opiate activity, and the Gly^3-Phe^4 sequence of enkephalins is inverted (Phe^3-Gly^4).

CONCLUSIONS

Conventional opioid peptides would no longer be considered as the only endogenous ligands for opioid receptors, and several conclusions concerning enkephalins an endorphins would have to be reapproached in the light of the occurrence of another distinct, previously ignored set of natural ligands: the dermorphins.

However, the dermorphin-like immunoreactivity present in mammalian tissues need to be characterized further and the peptide content isolated and prepared in highly purified form in order to determine the complete sequence. If mammalian dermorphin-like immunoreactivity differs from amphibian dermorphin, it may be necessary to immunize animals with this material in order to obtain antibodies of sufficient avidity and specificity for radioimmunoassay and immunocytochemistry of mammalian tissues.

We have emphasized already the unusual presence of a D-amino acid residue in dermorphin molecule. Immunochemistry demonstrated that this "biochemical aberration" also exists in the mammalian dermorphin. From a genetic point of view, this is an intriguing problem!

ACKNOWLEDGEMENT

Original research quoted in this paper has been supported throughout by a grant from Ministero Pubblica Istruzione, Ricerche

di Ateneo, Progetto "Fenomeni di membrana", University " La
Sapienza", Rome, Italy.

REFERENCES

Aloisi, F., Passarelli, F., Scotti de Carolis, A. and Longo, V.G.
    (1982), Central effects of dermorphin . Ann. Ist.Sup. Sanità
    (Roma) 18, 1-6.
Beddell, C.R., Clark, R.B., Hardy, G.W., Lowe, L.A., Ubatuba, F.B.
    Vane J.R., Wilkinson, F.R.S., Wilkinson, S., Chang, K.J.,
Cuatrecasas, P., and Miller, R.J. (1977). Structural requirements
    for opioid activity of analogues of the enkephalins. Proc.
    R. Soc. Ser. B, 198, 249-265.
Broccardo, M., Erspamer, V., Falconieri Erspamer, G., Improta, G.,
    Linari, G., Melchiorri, P., and Montecucchi, P.C. (1981).
    Pharmacological data on dermorphins, a new class of potent
    opioid peptides from amphibian skin. Br. J. Pharmacol., 73,
    625-631.
Broccardo, M., Improta, G., Nargi, M., and Melchiorri, P. (1982).
    Effects of dermorphin on gastrointestinal transit in rats.
    Regul. peptides, 4, 91-96.
Broccardo, M., Improta, G., Negri, L., and Melchiorri, P. (1983)
    Tolerance and physical dependence to dermorphin in rats.
    Peptides, in press.
Buffa, R., Solcia, E., Magnoni, E., Rindi, G., Negri, L., and
    Melchiorri, P., (1982). Immunohistochemical demonstration of
    a dermorphin-like peptide in the rat brain. Histochemistry,
    76, 273-276.
De Castiglione, R., Faoro, F., Perseo, G., Piani, S., Santangelo,
    P., Melchiorri, P., Falconieri Erspamer, G., Erspamer, V.,
    and Guglietta, A. (1981). Synthetic peptides related to the
    dermorphins. I. Synthesis and biological activity of the
    shorter homologues and analogues of the heptapeptide.
    Peptides , 2, 265-269.
Doble, K.E. and Greenberg, M.J. (1982). The clam is sensitive to
    FMRFamide, the enkephalins and their common analogs.
    Neuropeptides 2, 157-167.
Gorin, F.A., Balasubramanian, T.M., Cicero, T.J., Schwietzer, J.,
    and Marshall, G.R., (1980). Novel analogues of enkephalin:
    identification of functional groups required for biological
    activity . J. Mednl. Chem. 23, 1113-1122.

Gullner, H.G. and Kelly, G.D. (1983); Dermorphin: effects on
    anterior pituitary function in the rat. Arch. Int.
    Pharmacodyn., 256, 4-9.

Improta, G., Broccardo, M., Lisi, A. and Melchiorri, P., (1982).
    Neural regulation of gastric acid secretion in rats:
    influence of dermorphin. Regul. peptides 3, 251-256.

Kuhar, M.J., (1978). Histochemical localization of opiate receptor
    and opioid peptides. Fed. proc., 37, 153-157.

Lezoche, E., De Pasuale, G., Carlei, F., Procacciante, F.,
    Mariani, P., Nigri, R., Luminari, M., Speranza, V. (1983).
    Effect of dermorphin, a new opiate-like peptide, on gastric
    secretion, exocrine pancreatic secretion and gall bladder
    motility in man. Regul. peptides. Suppl. 2, 139-140.

Melchiorri, P., and Guglietta, A., (1983). Dermorphin: central
    sites of analgesia, catalepsy and inhibition of gastric
    secretion and gastric emptying. Regul. peptides, Suppl. 2,
    110.

Montecucchi, P.C., De Castiglione, R., Piani, S., Gozzini, L.,
    Erspamer, V. (1981a). Amino acid composition and sequence of
    dermorphin, a novel opiate-like peptide from the skin of
    Phyllomedusa sauvagei. Int. J. Peptide Protein Res. 17,
    275-238.

Montecucchi, P.C., De Castiglione, R. and Erspamer, V. (1981b).
    Identification of dermorphin and Hyp[6]-dermorphin in skin
    extracts of the Brazilian frog Phyllomedusa rhodei. Int. J.
    Peptide Protein Res., 17, 316-321.

Motta, M., Giudici, D., D'Urso, R., Falaschi, P., Negri, L. and
    Melchiorri, P. (1983). Dermorphin stimulates prolactin secr-
    etion in the rat. Neuroendocrinology, in press.

Negri, L., Melchiorri, P., Falconieri Erspamer, G., and Erspamer,
    V., (1981). Radioimmunoassay of dermorphin-like peptides in
    mammalian tissues. Peptides, 2, Suppl. 2, 45-49.

Negri, L., and Improta, G. (1983). Distribution and metabolism of
    dermorphin in rats. Pharmacol. Res. Commun., in press.

Nisticò, G., De Sarro, G.B., Rotiroti, D., Melchiorri, P. and
    Erspamer, V., (1981). Comparative effects of B-endorphin and
    dermorphin on behaviour, electrocortical activity and
    spectrum power in chicks after intraventricular
    administration. Res. Commun. Psychol. Psychiat. Behav., 6,
    315-364.

Puglisi-Allegra, S., Castellano, C., Filibeck, U., Oliviero, A.,
    and Melchiorri, P. (1982). Behavioural data on dermorphin in
    mice. Eur. J. Pharmacol., 82, 223-227.

GENETIC ANALYSIS OF CYCLIC AMP RESPONSE IN CULTURED CELLS

Michael M. Gottesman, George Vlahakis and Charles Roth

Laboratory of Molecular Biology
National Cancer Institute
National Institutes of Health
Bethesda, Maryland U.S.A. 20205

INTRODUCTION

The action of many peptide hormones has been shown to be
mediated by cyclic AMP acting as a second messenger to transmit the
effect of the hormone from the plasma membrane to the interior of
the cell[1]. It has been proposed that the only action of cyclic AMP
within cells is to stimulate the activity of the enzyme cyclic AMP
dependent protein kinase[2]. This enzyme activity is present within
cells in two forms, known as Type I and Type II cyclic AMP dependent
protein kinase (PKI and PKII, respectively)[2]. These two kinases
have the same catalytic subunit, but have different regulatory subunits. Cyclic AMP activates this tetrameric enzyme by binding to
the two regulatory subunits, which results in release of two active
catalytic subunits. In recent years, as a result of genetic
analysis of cultured cells in which mutants resistant to cyclic
AMP effects have been isolated, it has become apparent that PKI
and PKII are both involved in the mediation of cyclic AMP effects.
No genetic data has been obtained to date to prove that there are
any cyclic AMP effects mediated other than by the action of PK's
(reviewed in 3).

Although PK has been implicated in cyclic AMP action in many
cells types, in most cases the precise substrates which are phosphorylated by this enzyme have not been determined. This is
especially the case with respect to the effects of cyclic AMP on
cultured cells. Cyclic AMP treatment of cultured cells can be
effected by addition of cyclic AMP analogs or use of agents which
raise cyclic AMP levels in cells, such as cholera toxin (stimulates
adenylate cyclase) or methylisobutylxanthine (inhibits phosphodi-

esterase. These treatments result in a variable pleiotypic response. For example, in some cells such as epidermal cells[4], Swiss 3T3 cells[5], melanocytes[6] and mammary cells[7], growth is stimulated by cyclic AMP, whereas in other cell types, such as fibroblasts[8,9], adrenal Y-1 cells[10] and lymphoid cells[11] growth is inhibited. Other effects of cyclic AMP action include change in cell shape[8,9], altered cell adhesiveness and agglutinability[12-14], reduced sugar and amino acid transport[15], stimulation of gap junction formation[16] and increased activity of various enzymes such as transglutaminase[17], ornithine decarboxylase[18], and cyclic nucleotide phosphodiesterase[19]. In each of these cases, PK has been implicated genetically as the mediator of the cyclic AMP effect, but in no case have the substrate or substrates for this enzyme been biochemically defined. In this paper, we will review work from our laboratory which utilizes a genetic analysis of Chinese hamster ovary (CHO) cells to prove the role of cyclic AMP dependent PK in mediation of the growth inhibition of these cultured fibroblasts. We have also found that when CHO cells are transformed by Rous sarcoma virus (RSV), an avian RNA tumor virus, their growth is no longer inhibited by cyclic AMP, but may actually be stimulated by it[20,21]. The development of these two models, i.e., the CHO fibroblast, and the CHO cell transformed by RSV (CHO-RSV), has enabled us to define several cellular substrates for PK which might be involved in modulation of the growth phenotype of cultured cells by cyclic AMP.

GENETIC ANALYSIS OF CYCLIC AMP-INDUCED GROWTH INHIBITION

The growth of CHO cells is inhibited by cyclic AMP, but the cells are not killed and simply grow at a reduced rate. To isolate mutants of CHO cells resistant to cyclic AMP, we exposed mutagenized cells to 8-Br-cyclic AMP and suspended them in agar. Resistant clones are much larger than sensitive clones under these conditions and appear at a frequency of approximately $1/10^5$ cells plated. About 50% of these clones appear to breed true with respect to cyclic AMP resistance. Over the past several years, we have analyzed 30 such independent mutants biochemically, and all but perhaps 1 or 2 appear to have defective PK's. We have not been able to determine the defect in the mutants with apparently normal PK's, and it is conceivable that they have defective PK's in which the alteration is too subtle for us to detect, or other alterations involving phosphatases or protein substrates for the PK's.

Eight mutants have been analyzed in detail by DEAE chromatography to determine the presence or absence of the two PK forms, PKI and PKII. On the basis of this analysis, we have divided

our mutants into two general classes: Class I lacks PKII either
as a result of an alteration in the catalytic subunit[22], the
Type I regulatory subunit[23] or other undefined alterations.
Class I mutants are all dominant in somatic cell hybrids, and
recently we have found that the biochemical defects in these
mutants are transmissible by DNA-mediated gene transfer[24].
Class II mutants lack PKI, and have variable amounts of PKII.
These mutants are genetically recessive and have been found to
fall into two complementation groups[25]. Mutants in one of these
complementation groups have greatly reduced levels of the catalytic
subunit of PK[26]. We have not characterized any mutants to date
which completely lack PK activity.

   We have used these mutants to define substrates which might
be involved in mediation of cyclic AMP-induced growth inhibition.
Cells were labeled with ^{32}P-orthophosphate and Class I mutants,
Class II mutants and wild-type cells were either treated with 8-
Br-cyclic AMP or were untreated. The results of these experiments[27]
indicated that in the Class I mutants (no PKII) two intracellular
protein kinase substrates including a 50K protein of unknown function [27], and the intermediate filament protein, vimentin[23] were
not phosphorylated after cyclic AMP treatment in the same manner
as they were in wild type cells. In addition, Class I mutants
have been found to be more sensitive to antimicrotubule agents,
such as Colcemid, and some of these Colcemid-resistant lines
derived from Class I mutants appear to be reverted for their PK
defect[28]. This genetic observation argues that PKII also phos-
phorylates a substrate involved in microtubule polymerization,
such as perhaps a microtubule associated protein. Although no
phosphorylation defect has yet been found in Class II mutants
(no PKI), these mutants are equally resistant to cyclic AMP effects
on growth, and it is assumed that our methods for detecting phos-
phoprotein substrates are simply not sensitive enough to detect
PKI substrates.

   One way in which these PK mutants can be utilized is to
determine whether drugs, such as interferon and tumor promoters
which are known to raise intracellular cyclic AMP levels, actually
exert their cellular effects through a mechanism involving PK.
For example, we have recently found that human β-interferon
inhibits CHO cell growth and proliferation of encephalomyocarditis
virus. These interferon effects occur in CHO PK mutants[29] indicating that they are not mediated by PK. Similarly, although
ornithine decarboxylase induction by cyclic AMP is blocked in PK
mutants, this enzyme is induced by the tumor promoter TPA in both
wild-type CHO cells and PK mutants[18].

   The studies described above are summarized in Table I.

Table 1.   Characterization of Cyclic AMP Resistant CHO Mutants

Property	Class I	Class II
PK phenotype	PKII absent PKI present	PKI absent PKII reduced
Number of mutants	5	3
Defects in PK subunits	1. Altered C (1) 2. Altered RI (1) 3. Unknown (3)	1. Reduced levels of C (2) 2. Unknown (1)
Genetic dominance	Dominant	Recessive
DNA-mediated trans- fer of phenotype	Yes	No
Phosphorylation of 50K protein, vimentin	No	Yes
Defective in cAMP- induced growth inhi- bition, morphological change, glucose and amino acid transport, transglutaminase, ornithine decarboxylase and phosphodiesterase induction	Yes	Yes
TPA induces ornithine decarboxylase	Yes	Yes
Human β-interferon inhibits growth and has anti-viral properties	Yes	Yes

ANALYSIS OF GROWTH STIMULATION BY CYCLIC AMP

One of the difficulties in studying growth inhibition of CHO cells by cyclic AMP is that the mechanism which regulates growth in wild-type CHO cells is not understood. In order to produce a more defined system for the study of cyclic AMP effects on cell growth, we transformed cells with RSV. This virus has been extensively studied genetically and biochemically, and it has been established that pp60^{src}, the product of the viral src gene, is both necessary and sufficient to transform mammalian cells[30]. Since CHO cells are already transformed with respect to tissue culture criteria (i.e., growth in suspension, growth in low serum and focus-formation), it was necessary to use a line of CHO cells with more "normal" growth properties. For these studies, we started with a CHO cell line developed by Pollard[31] which would not grow in agar. We were able to isolate several independent RSV transformants of this cell line by selecting for growth in agar. Evidence that these lines are truly transformed by RSV, as opposed to being spontaneous transformants derived from the parental non-transformed CHO line is as follows: 1) The frequency of CHO-RSV is approximately 10^{-4}, as opposed to 10^{-6} to 10^{-5} for spontaneous transformants; 2) CHO-RSV cells contain the RSV virion as indicated by Southern blotting and virus rescue experiments; 3) We have isolated flat, nontransformed revertants of CHO-RSV which have either lost the entire RSV virion, or have specific defects in the RSV src gene; and (4) We have isolated CHO-RSV in which transformation is temperature-sensitive as a result of using an RSV encoding a temperature-sensitive src product.

As shown in Table 2, CHO-RSV cells are poorly tumorigenic in nude female mice (line 2), whereas the parental, spontaneously transformed CHO cells are highly tumorigenic (line 1). When we treated both of these cell types with cholera toxin prior to inoculating the cells in the nude mice, we found that, as expected, the ability of the parental cells to form tumors was inhibited (line 3), but the ability of the CHO-RSV to form tumors was dramatically stimulated (line 4). These results argue that cyclic AMP has diametrically opposed effects on growth in spontaneously transformed cells as in cells transformed by RSV, presumably because growth is regulated in very different ways in these two cell types.

Table 2. Tumorigenicity of CHO Cells

Cell Type[a]	Treatment[b]	Tumor formation[c]
1. CHO	None	60%(6/10)
2. CHO-RSV	None	8%(3/38)
3. CHO	Cholera toxin	20%(1/5)
4. CHO-RSV	Cholera toxin	46%(18/39)

[a] 3×10^5 CHO cells or 1×10^7 CHO-RSV cells were inoculated subcutaneously into female Swiss nude (nu/nu) mice which were treated 3 times per week with a mixture of anti-lymphocyte and anti-thymocyte serum (M.A. Bioproducts).
[b] Untreated cells were washed with PBS three times prior to inoculation; treated cells were exposed to cholera toxin, 100 ng/ml for 30 minutes, and then washed three times in PBS.
[c] Mice were examined for tumors twice a week for 6 weeks. Appearance of tumors greater than 5 mm in diameter after 3 weeks was considered a positive response.

We have examined the histology of the CHO-RSV tumors induced by cholera toxin and find that they are composed of basophilic nests of round and fusiform cells, which is very similar to the appearance of these cells in tissue culture. An example of a low and high-power field of a histological section is shown in Figure 1. The tumors are highly vascular and seem to be encapsulated, as is usually seen in tumors growing from heterologous cells in nude mice.

To be certain that these cells were, indeed, CHO cells, as opposed to mouse cells stimulated to grow into tumors either by the cholera toxin treatment or the CHO cell inoculum, tumor-derived cells were grown in culture. These cultured cells were karyotyped by standard methods and their karyotypes (shown in Figure 2) were compared with that of the parental RSV-CHO cells which were originally inoculated into the mice. The morphology and modal number of chromosomes is identical in the parental CHO-RSV cell line and in the two cell lines we examined which derived from tumor material. We were also able to rule out the possibility that cholera toxin treatment and tumor formation selected a small population of CHO-RSV cells which were tumorigenic, by showing that the two cell lines derived from the tumors were not tumorigenic unless they were pre-treated with cholera toxin[21].

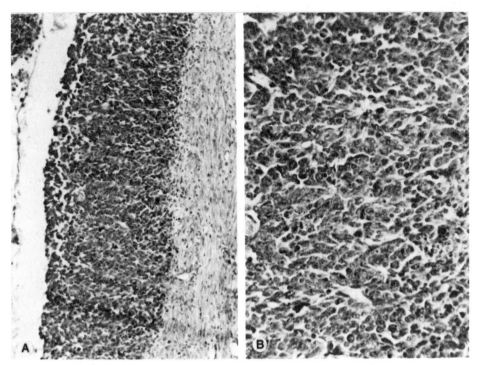

Figure 1. Histological analysis of tumors formed in nude mice by CHO-RSV. A) low power field showing encapsulation of tumor, B) high power field.

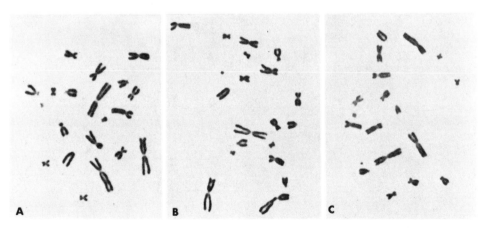

Figure 2. Karyotypic analysis of CHO-RSV tumor cells. A) parental CHO-RSV line, B) CHO-RSV strain 1681 derived from tumor, C) CHO-RSV strain 1690 derived from tumor.

What is the mechanism of this stimulation of tumorigenicity
of CHO-RSV by cyclic AMP?   To examine this question, it was necessary to study the effect of cyclic AMP treatment on pp60<u>src</u>,
the product of the viral transforming gene which is known to control growth in RSV transformed cells.   Collett et al.[31] have previously observed that in a crude <u>in vitro</u> system, pp60<u>src</u> was a
substrate for cyclic AMP dependent protein kinase, and that a
serine in the amino-terminal portion of the molecule was phosphorylated in this reaction.   We determined that this phosphorylation
also occurred in intact cells by treating CHO-RSV with cyclic AMP
and labelling with ^{32}P-orthophosphate.   The pp60<u>src</u> was immunopre-
cipitated with a specific antiserum and one and two-dimensional
peptide mapping using <u>Staphylococcus</u> V8 protease and trypsin was
performed on the purified pp60<u>src</u>.   From these analyses, it
was possible to determine the time course of the appearance of
phosphoserine in the amino-terminus of the pp60<u>src</u> molecule after
cyclic AMP treatment.   These data are shown in Figure 3.   Also
shown in this figure is the time course of appearance of phosphotyrosine in the carboxy-terminus of pp60<u>src</u>. As can be seen,
phosphoserine content increases up to six-fold after 2-3 hours
of cyclic AMP treatment, whereas phosphotyrosine content (presumably a measure of autophosphorylation) does not significantly
change over time.

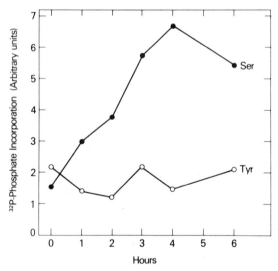

Figure 3.   Presence of Ser-P and Tyr-P in pp60<u>src</u> as a function of
          duration of cyclic AMP treatment.

   We next examined the effect of this phosphorylation of
pp60[src] on its activity. It has been known for some time that
pp60[src] is a tyrosine kinase[33]. This tyrosine phosphotransferase
activity was measured in extracts of CHO-RSV cells either treated
or not treated with 8-Br-cyclic AMP or cholera toxin. The results
of one such analysis are shown in Table 3. In several experiments
we observed a consistent 2-3 fold stimulation of the tyrosine
phosphotransferase activity of pp60[src] as measured by phosphorylation of the heavy chain of IgG used in the immune complex kinase
kinase assay. This stimulation was observed in both supernatant
and pellet fractions. The time course of this stimulation of
activity has also been shown to correlate with the appearance of
phosphoserine in the pp60[src] molecule[34]. These results are consistent with the idea that stimulation of tumorigenicity of
CHO-RSV cells is the result of activation of the phosphotyrosine
kinase activity of pp60[src] by cyclic AMP dependent phosphorylation
of this molecule.

Table 3.   Activation of pp60[src] kinase activity by cyclic
           AMP treatment of CHO-RSV

| | Tyrosine Kinase Activity, cpm[a] | |
	No Treatment[b]	1mM 8-Br-cAMP[b]
Soluble Fraction	35,000	107,000
Pellet Fraction	100,000	195,000

[a] pp60[src] kinase activity was measured as the transfer of [32]P
from 8-^{32}P-ATP to the heavy chain of anti-pp60[src] IgG. The heavy
chain was cut out of SDS-PAGE gels and incorporated radioactivity
was determined by liquid scintillation counting as previously
described[34].
[b] CHO-RSV cells were either treated with 1mM 8-BrcAMP for 3h
or were not treated. Extracts were prepared by sonication and
supernatant and pellet fractions were obtained as previously
described[34].

SUMMARY AND CONCLUSIONS

We have developed two systems for the study of cyclic AMP effects on cell growth: the CHO cell, whose growth is inhibited by cyclic AMP, and the CHO-RSV cell, whose growth may be stimulated by cyclic AMP. Analysis of mutant CHO cell lines selected for resistance to cyclic AMP has allowed us to formulate the following major conclusions: 1) PK mediates the action of all known cyclic AMP effects on CHO cells; 2) mutants with altered PK's are fully viable; 3) both PKI and PKII are needed to mediate cyclic AMP effects, but only substrates for PKII have been defined using our mutants; 4) PKII substrates within intact cells include a 50K protein of unknown function, vimentin, and probably a component of the microtubule system; and 5) inhibition of cell growth by cyclic AMP appears to be a complex process involving phosphorylation of many protein substrates by both PKI and PKII.

Analysis of the CHO-RSV system has revealed that growth stimulation of these cells by cyclic AMP correlates with phosphorylation of $pp60^{src}$ and activation of its tyrosine kinase activity. One prediction from these studies is that for cell lines such as epidermal cells, which are stimulated by cyclic AMP, growth may be controlled by a cellular analog of src or another growth-promoting protein similarly regulated by cyclic AMP. These results also suggest that growth-promoting pathways may differ considerably from tissue to tissue in their responsiveness to cyclic AMP and this fact should be taken into account in the development of tumor chemotherapy.

ACKNOWLEDGEMENTS

We would like to thank Deborah Chertock for typing this manuscript, Ray Steinberg for photographic assistance and Irene Abraham for critical comments on the manuscript. Special thanks go to our many colleagues in the Laboratory of Molecular Biology, National Cancer Institute, for ideas and scientific contributions to this work and to Ira Pastan for his invaluable advice and support.

REFERENCES

1. Robison, G.A., Butcher, R.W., and Sutherland, E.W.; Cyclic
    AMP (Academic Press, New York); 1971   .

2. Krebs, E.G., and Beavo, J.A.; Annual Review of Biochemistry;
    48: 923-959; 1979

3. Gottesman, M.M.; Cell; 22:329-330; 1980 4. Green, H.; Cell;
    15:801-811; 1978 5. Pruss, R.M. and Herschman, H.R.; J. Cell.
    Physiol.; 98:469-474; 1979

6. Eisinger, M. and Marko, O.; Proc. Natl. Acad. Sci. U.S.A.;
    79:2018-2022; 1982

7. Taylor-Papadimitrou, J., Purkis, P. and Fentiman, I.S.; J.
    Cell. Physiol.; 102:317-321; 1980

8. Johnson, G.S., Friedman, R.M. and Pastan, I.; Proc. Natl.
    Sci. U.S.A.; 68:425-429; 1971

9. Hsie, A.W. and Puck, T.T.; Proc. Natl. Acad. Sci. U.S.A.;
    68:358-361; 1971

10. Gutmann, N.S., Rae, P.A. and Schimmer, B.P.; J. Cell. Physiol.;
    97:451-460; 1978

11. Hochman, J., Insel, P.A., Bourne, H.R., Coffino, P. and
    Tomkins, G.M.; Proc. Natl. Acad. Sci. U.S.A.; 72:5051-5055;
    1974

12. Johnson, G.S. and Pastan, I.; J. Natl Cancer Inst.; 48:1377-
    1387: 1972

13. Willingham, M.C. and Pastan, I.; J. Cell Biol.; 63:288-294; 1974

14. Otten, J., Johnson, G.S. and Pastan, I.; Biochem. Biophys.
    Res. Commun.; 44:1192-1198; 1971

15. LeCam, A., Gottesman, M.M., and Pastan, I.; J. Biol. Chem.;
    255:8103-8108; 1980

16. Wiener, E.C. and Loewenstein, W.R.; Nature; 305:433-435; 1983

17. Milhaud, P.G., Davies, P.J.A., Pastan, I. and Gottesman,
    M.M.; Biochem. Biophys. Acta; 630:476-484; 1980

18. Lichti, U. and Gottesman, M.M.; J. Cell. Physiol.; 113:433-
    439; 1982

19. Bourne, H.R., Tomkins, G.M. and Dion, S.; Science; 181:952-954; 1973

20. Roth, C.W., Singh, T.J., Pastan, I and Gottesman, M.M.; J. Cell. Physiol.; 111:42-48; 1982

21. Gottesman, M.M., Roth, C., Vlahakis, G. and Pastan, I.; submitted for publication; 1983

22. Evain, D., Gottesman, M.M., Pastan, I. and Anderson, W.B.; J. Biol. Chem.; 254:6931-6937; 1979

23. Gottesman, M.M., Singh, T.J., Le Cam, A., Roth, C., Nicolas, J.C., Cabral, F. and Pastan, I.; Cold Spring Harbor Conferences on Cellular Proliferation; 8:195-209; 1981

24. Abraham, I., Brill, S. and Gottesman, M.M.; in preparation

25. Singh, T.J., Roth, C., Gottesman, M.M., Pastan, I.H.; J. Biol. Chem.; 256:926-932; 1981

26. Murtaugh, M.P., Steiner, A.L. and Davies, P.J.A.; J. Cell Biol; 95:64-72; 1982

27. Le Cam, A., Nicolas, J.-C., Singh, T.J., Cabral, F., Pastan, I. and Gottesman, M.M.; J. Biol. Chem.; 256:933-941; 1981

28. Abraham, I., Hyde, J. and Gottesman, M.M.; in preparation

29. Banerjee, D.K., Baksi, K. and Gottesman, M.M.; Virology; 129:230-238; 1983

30. Bishop, J.M. and Varmus, J.; in RNA Tumor Viruses; eds.-Weiss, R., Teich, N., Varmus, H. and Coffin, J.; Cold Spring Harbor Laboratory; 999-1108; 1982

31. Pollard, J.W. and Stanners, C.P.; J. Cell Physiol.; 98:571-586; 1979

32. Collett, M.S., Erikson, E. and Erikson, R.L.; J. Virol.; 29:770-781; 1979

33. Collett, M.S. and Erikson, R.L.; Proc. Natl. Acad. Sci. U.S.A.; 75:2021-2024; 1978

34. Roth, C., Richert, N.D., Pastan, I. and Gottesman, M.M.; J. Biol. Chem.; 258:10768-10773; 1983

GROWTH REGULATION IN S49 MOUSE LYMPHOMA: INVOLVEMENT OF CYCLIC AMP

AND SUBSTRATE ADHESIVENESS

Jacob Hochman and Efrat Levy

Department of Zoology*
The Hebrew University of Jerusalem
Jerusalem, Israel

INTRODUCTION

We have recently become interested in various aspects of growth regulation in lymphoma cells[1-4], our model system being the Balb/C derived T-cell S49 lymphoma[5]. S49 cells have mainly been used to study hormone regulated pathways leading to growth arrest and cellular death.[6] These studies have all been carried out in vitro in cells growing in suspension culture. In this paper we report some of our recent findings in S49 cells pertinent to the scope of the symposium, namely: 1) The involvement of cyclic AMP and cyclic AMP-dependent protein kinase in the in vivo growth of S49 tumors in nude mice. 2) The effect of substrate adhesiveness on the in vitro response to cyclic AMP. 3) The relationship between substrate adhesiveness and progressive tumor growth in syngeneic Balb/C mice.

METHODS

Cells

S49 cells and the various adherent variants derived from them were maintained in Dulbecco's modified Eagle's medium supplemented with 10% heat inactivated serum. This was carried out at 37°C in a humidified atmosphere containing 5% CO_2.

Protein Kinase Assay

Cyclic AMP-dependent protein kinase activity in cell extracts was carried out by measuring the transfer of ^{32}P from ($\gamma ^{32}P$)-ATP

---

*Present address: Department of Genetics, The Hebrew University

155

(New England Nuclear) to histone in the absence and in the presence of increasing cyclic AMP concentrations as previously described.[7]

Cell Attachment Assay

Prior to the experiment adhering cells were detached from their substratum by pipetting. Cells were adjusted to 1×10^6 cells per ml and portions of 10 ml were added to 50 ml tissue culture flasks (Costar, 25 cm^2 Cat. No 3050) and incubated at 37°C. At the time points indicated flasks were tilted, aliquots were taken out and the number of cells remaining unattached was scored. From this, the percentage of attached cells was calculated. Assays were carried out only on samples in which viability was greater than 95%.

Fluorescence Activated Cell Sorter (FACS) Analysis

Cell sorter analysis was carried out on sera derived from immunized mice and Balb/C controls. Both sera were incubated with suspension growing as well as adherent cell populations (45 min. at 4°C) extensively washed with Hank's solution (without phenol red) and incubated with FITC-conjugated goat anti-mouse IgG Fc fragment (cappel) for 45 min. at 4°C. Samples were analyzed with the FACS-II cell sorter (Becton & Dickinson).

RESULTS AND DISCUSSION

I. Tumor Growth and Cyclic AMP Resistance

In S49 cells incubation in vitro with DiButyryl cyclic AMP (DBcAMP) or agents that increase intracellular concentrations of cyclic AMP results in growth arrest (at the G1 phase of the cell cycle) followed by cell death within 48-72 hours.[8] These findings initiated a combined biochemical-genetic approach based on the isolation and molecular characterization of various classes of mutants resistant to DBcAMP and hormones that activate adenylate cyclase. Thus, mutants with different lesions in either the cell membrane bound adenylate cyclase complex, or the cyclic AMP-dependent protein kinase, or a site distal to protein kinase activation have been characterized.[9-13] Of particular interest in the present context are the mutant cells that are devoid of cyclic AMP-dependent protein kinase activity, therefore being resistant to exogenous DBcAMP. The findings that these mutant cells grow in suspension culture in a perfectly normal manner led to the conclusion[14] that in S49 cells cyclic AMP is a nonessential regulator of the cell cycle in vitro.

In our laboratory we have adapted a subline of mutant S49 cells (descendants of clone 24.6.1) that lack cyclic AMP-dependent protein kinase activity[9,10] (designated PK$^-$) to grow either as a solid or an ascitic form in congenitally athymic (nude) mice.[1] Later we adapted a subline of parental S49 cells (descendants of clone 24.3.2) that demonstrate cyclic AMP-dependent protein kinase activity (designated PK^+) to grow as progressive tumors in nude mice. Throughout the initial in vivo passages of PK^+ and PK^- cells, before extensive selection for increased tumorigenicity could have taken place, it was found that PK^+ cells were less tumorigenic than the PK^- mutants derived from them (Fig. 1). Thus at each inoculum tested, a higher percentage of recipient nude mice developed progressive tumors when inoculated with PK^- cells than following inoculation with PK^+ cells. This difference in tumorigenicity becomes more pronounced at a lower inoculum of PK^+ and PK^- cells. Moreover, in those mice that developed tumors, tumors were detected earlier in PK^- than in PK^+ inoculated recipients.

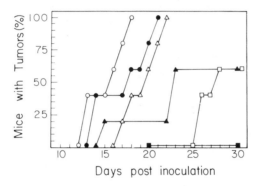

Fig. 1. Time of tumor appearance at different doses of PK^+ and PK^- cells. Male nude mice, 8-10 weeks old were inoculated intraperitoneally (n=5). Results show the first day at which a tumor could be detected. Open symbols - PK^- cells: (O–O) $1x10^7$/mouse, (△–△) $5x10^6$/mouse and (□–□) $2.5x10^5$/mouse; closed symbols - PK^+ cells: (●–●) $1x10^7$/mouse, (▲–▲) $5x10^6$/mouse and (■–■) $2.5x10^5$/mouse.

When PK^+ and PK^- cells ($2x10^7$/mouse) were investigated as to their ability to respond in vivo (in the mouse) to DiButyryl cyclic AMP (DBcAMP) it was found (Fig. 2): 1. Even at this high inoculum PK^+ tumors were detected later than PK^- tumors - 100% within 22 days as compared with 100% within 12 days, respectively. 2. DBcAMP delayed for a few days the appearance of PK^+ tumors while being ineffective in this respect upon PK^- tumors. Preliminary studies

with different schedules of DBcAMP treatment (data not shown) made
it possible to delay tumor development for longer periods of time,
and even inhibit it completely in some cases. Taken together these
findings suggest that host factors, through increase of intracellular cyclic AMP levels and subsequent activation of cyclic AMP-
dependent protein kinase, can delay in vivo growth of S49 cells.
Depletion of this system (as in PK⁻ cells) renders these cells less
responsive to growth inhibiting stimuli from the host, thereby
increasing their tumorigenic potential. In this respect it would
be of interest to find out whether S49 cells (or any other cell that
responds in a similar manner thereof) devoid of more than one
in vivo responding system, like a double mutant deficient in both
cyclic AMP-dependent protein kinase activity and glucocorticoid
responsiveness (glucocorticoids are growth inhibiting to S49 cells
as well) will demonstrate a still greater tumorigenic potential.

Various observations have suggested an inverse correlation
between cyclic AMP-dependent protein kinase and malignancy.[15-19]
Through the use of wild type (PK⁺) and kinase negative mutants
(PK⁻) of S49 cells our findings demonstrate a direct relationship
between the regulation of this enzyme and the in vivo control of
tumor growth.

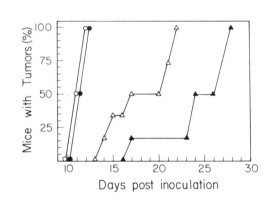

Fig. 2. Effect of DiButyryl cyclic AMP on development of PK⁺ and
PK⁻ tumors. Male nude mice, 8-10 weeks old were inoculated intraperitoneally with 2×10^7 cells (n=6). DBcAMP-3mg/mouse/day was
injected for 5 consecutive days following cell inoculation.
(O-O) - PK⁻ controls, (△-△) - PK⁺ controls, (●-●) - PK⁻ +
DBcAMP, (▲-▲) - PK⁺ + DBcAMP. Results show the first day at
which a tumor could be detected.

II. Cyclic AMP Resistance and Cell Adhesiveness

From parental S49 cells that grow in suspension culture we have selected for and isolated spontaneous variants characterized by their ability to adhere to their substrate–bottom of tissue culture flask, and grow while attached to it. The selection and properties of these variants (designated – S49–Adh) have been previously described.[2] They grow attached to their substratum until a dense monolayer is formed. At that stage, daughter cells are being released and grow as a suspension borne population without loss of substrate adhesiveness. Upon transfer of the daughter cells to a new flask they will adhere to the substrate with the same kinetics as their parental cells. The adherent phenotype is a stable trait and S49–Adh cells do not lose it even after extended periods of in vitro growth in culture.

From S49–Adh cells we selected for clones resistant to 2×10^{-4}M DBcAMP. To that effect S49–Adh cells were cloned in semi–solid agarose[20] in the presence of the drug. From 20 tissue culture dishes (1×10^6 cells/dish) two resistant clones, designated DC_1 and DC_2, were isolated. These were picked up, transferred into tissue culture flasks and were grown to mass culture. Both clones remained adherent to their substrate. Fig. 3 demonstrates the substrate adhesion kinetics of clone DC_1 as compared with suspension growing S49 cells. This curve is similar to our previously reported data in S49–Adh cells.[2] When DC_1 cells were grown continuously in culture we have

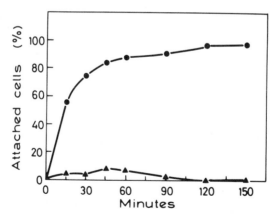

Fig. 3. Adhesion kinetics of adherent S49 variants (DC_1 cells) to their substrate. (●—●) – DC_1 cells, (▲—▲) – suspension growing ACT4 cells. Each time point represents the mean of duplicate assays.

observed (Fig. 4, C-F) that they gradually acquired resistance to increasing concentrations of exogenous DBcAMP. Thus, immediately following their isolation DC_1 cells were resistant to 2×10^{-4}M DBcAMP while being sensitive to 8×10^{-4}M DBcAMP (Fig. 4,C). Within a period of six months in culture, DC_1 cells became resistant to 1×10^{-3}M DBcAMP (Fig. 4,F). Suspension growing S49 cells (both PK^+ and PK^- sublines) that have been grown continuously as controls, together with DC_1 cells, have not undergone this gradual change in resistance: PK^+ cells (subline ACT4) remained sensitive to 2×10^{-4}M DBcAMP (Fig. 4,A) while PK^- cells (subline T-8) remained resistant to 1×10^{-3}M DBcAMP (Fig. 4,B). The genetic origin of this resistance is presently unclear – either a gradual increase in resistance that occurs concomitantly in the whole population, or a selective growth

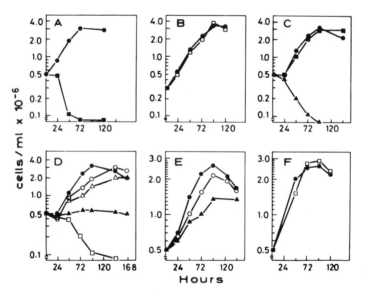

Fig. 4. Effect of DiButyryl cyclic AMP on growth of suspension growing and substrate adherent variants of S49 lymphoma cells. A – suspension growing (PK^+) ACT4 cells, B – suspension growing (PK^-) T-8 cells, C-F – Adherent DC_1 cells at various times following their cloning: 2 weeks, 3 months, 5 months and 6 months, respectively. Concentrations of DBcAMP: (●–●) – No drug, (■–■) – 0.2 mM, (O–O) – 0.4 mM, (△–△) – 0.6 mM, (▲–▲) – 0.8 mM and (□–□) – 1.0 mM, respectively.

advantage (over the whole population) in a small number of spontaneously appearing resistant variants. Nevertheless these findings
suggest that changes in the cell membrane related to substrate
adhesiveness (in as yet an unknown manner) can regulate the resistance to cyclic AMP in S49 lymphoma cells.

To study the molecular mechanism involved in cyclic–AMP resistance we tested the activity of cyclic AMP–dependent protein kinase
in both DC_1 and suspension growing S49 cells (PK^+ and PK^- sublines).
It was found (Fig. 5) that DC_1 cells demonstrate an activation curve
of cyclic AMP dependent protein kinase similar to that found in ACT4
cells, when tested on cellular extracts. Therefore, the resistance
of DC_1 cells to DBcAMP is not due to lack of enzyme activity (as the
case is in T-8 cells). In addition, the resistance of DC_1 cells to
1 mM DBcAMP does not stem from reduced cell membrane permeability to
the drug, insofar as incubation of intact DC_1 cells in culture medium
with 1 mM DBcAMP causes intracellular activation of cyclic AMP–dependent protein kinase very similar to that found in ACT4 cells when
carried out under identical conditions (data not shown).

The above mentioned findings suggest that the lesion in DC_1
cells might be either an inability of their cyclic AMP–dependent
protein kinase to phosphorylate the appropriate endogenous sub-
strate(s) or an inability of the substrate(s) themselves to undergo
phosphorylation by a normal enzyme. Alternatively, the lesion might
be at a site distal to subtrate phosphorylation by an activated

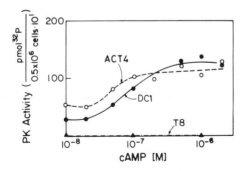

Fig. 5. Activation of cyclic AMP–dependent protein kinase from
suspension growing and substrate adhering variants of S49 cells.
Assays were performed on crude cellular extracts from ACT4, T-8 and
DC_1 sublines. Data shown are after substracting cyclic AMP independent kinase activity at each point.

kinase. If the later speculation is correct, then the lesion may
be located between substrate phosphorylation and growth arrest
(since in DC_1 cells, kinase activation is not followed by growth
arrest and cellular death). In that case, DC_1 cells can be considered as true growth ("G") variants, the study of which should
enable one to delineate the molecular pathways leading to growth
arrest and subsequent cell death in S49 cells in response to cyclic
AMP.

When DC_2 cells were propagated in culture and studied, in the
same manner as DC_1 cells, for their cyclic AMP resistance and protein
kinase activation, similar results were obtained.

III. Cell Adhesiveness and Tumor Growth

Cell adhesiveness - the ability of cells to adhere to other
cells and to their substrata is of fundamental importance in a
variety of biologically important phenomena including tumor development and metastasis.[21-24] The relationship between cell adhesiveness, transformation and metastasis has been studied extensively
in fibroblastoid cells that grow, in vitro, as monolayers attached
to their substratum and to each other.[25-28] Thus increased tumorigenicity in fibroblastoid cells is very often correlated with
decreased cell adhesiveness. We have undertaken to study this
relationship between cell adhesiveness and tumorigenicity in
malignant lymphoid cells that grow in suspension culture as free
floating cells normally devoid of cell-cell or cell-substrate interactions. Our working hypothesis was as follows: If the inverse
relationship between cell adhesiveness and tumorigenicity in cells
of the fibroblastoid category is generally more applicable, then
the selection of adhesive variants from free floating malignant
lymphoid cells should result in variants with decreased tumorigenic
potential.

Parental, suspension borne, S49 cells (from which the above
mentioned S49-Adh, DC_1, DC_2 and other adherent sublines were
derived) were relatively non-tumorigenic in syngeneic Balb/C mice.
This is attributed to changes these cells have undergone while
growing for extended periods of time in suspension culture. To
overcome this obstacle and to test our working hypothesis we first
readapted parental S49 cells (derived from various clones) to grow
as progressive tumors in Balb/C mice. Through continuous in vivo
passages for up to 40 consecutive generations we ended up with
highly tumorigenic S49 cells that developed progressive tumors with
an inoculum as low as 100-1000 cells per mouse (in preparation).
For routine transfers we used an inoculum of 10^7 cells per mouse
given intraperitoneally. Such an inoculum developed as a disting-
uishable tumor within 7-10 days and killed the host within 15 days.
Tumor cells were then readapted to grow in vitro in suspension

culture while retaining their tumorigenic potential. From one of
these sublines (T-40) we selected a substrate adhering variant named
T-40-Adh. When T-40 and T-40-Adh cells (as well as other adherent
variants derived in the same manner[3]) were compared as to their
ability to develop tumors in Balb/C mice, it was found (Fig. 6) that
the former cells developed progressive tumors whereas the later cells
were non-tumorigenic. These findings demonstrated, for the first
time, that, at least in one malignant lymphoma, our working hypothesis was valid. Moreover, following our original report[3] it has been
recently demonstrated[29] that a substrate adherent variant derived
from a highly metastatic suspension growing mouse lymphoma (L5178Y),
had a greatly decreased metastatic potential. These findings suggest
that our approach might be applicable to other malignant lymphoid
cells as well.

In addition to the impaired tumorigenic potential of adherent
S49 variants it was also found[3] that Balb/C mice inoculated with
these cells were immunized against a challenge of 10^7 suspension
growing, tumorigenic S49 cells. These findings in S49 cells suggest
a new and somewhat different approach (compared to previous attempts)
towards the xenogenization[30-31] of malignant lymphoid cells, thereby

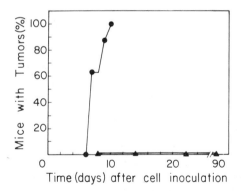

Fig. 6. Tumorigenicity of suspension growing S49 cells and their
adherent variants in syngeneic mice. Balb/C mice 8-10 weeks old
were inoculated intraperitoneally with 2×10^7 cells of either T-40
parental cells (●—●), or T-40-Adh variants (▲—▲) derived from
them. Mice were checked daily and the first day at which tumors
could be recognized was recorded.

rendering tumor cells antigenically foreign to their host so that
they may be used more effectively in immunotherapy and immuno-
diagnosis.

    To study the immune protective mechanisms involved we have used
various approaches (in preparation). One approach was to study the
serum of the immunized Balb/C mice. We found (Fig. 7) that mice
inoculated with 2×10^7 adherent cells, followed by a challenge of
2×10^7 suspension growing cells, revealed in their serum, antibodies
that were able to recognize both adherent and suspension growing S49
cells. These antibodies did not interact with spleen cells from
Balb/C mice, nor did normal mouse serum interact with either adherent
or suspension growing S49 cells. Antibody analysis was based on
cytofluorography using FiTC-conjugated second antibody as probe.
In recent, preliminary, experiments we have tried to identify antigenic determinants, in adherent variants, through immunoprecipitation

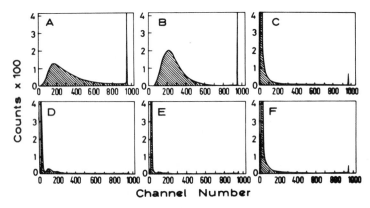

Fig. 7. Histograms from cytofluorgraphic analysis of the interaction
between adherent S49 cells and immune serum from syngeneic mice.
Immune serum was derived from Balb/C mice previously inoculated and
challenged with 2×10^7 cells of adherent and suspension growing S49
cells, respectively. A - Immune serum and adherent cells.
B - Immune serum and suspension growing cells. C - Immune serum
and spleen cells from Balb/C mice. D - Normal mouse serum and
adherent cells. E - Normal mouse serum and suspension growing cells.
F - Normal mouse serum and spleen cells from Balb/C mice.
The histograms demonstrate the number of cells stained at each
relative fluorescence level (channels 100-1000).

of ^{35}S-methionine labelled cellular extracts of these cells with
either serum from immunized mice or normal mice. Fig. 8 demonstrates
the results of one such experiment in which a series of sera from
individually bled immunized and control mice were used for immunoprecipitation of adherent cell extracts. This was followed by
electrophoresis and fluorography. Both qualitative as well as quantitative differences between the control and the immune sera could

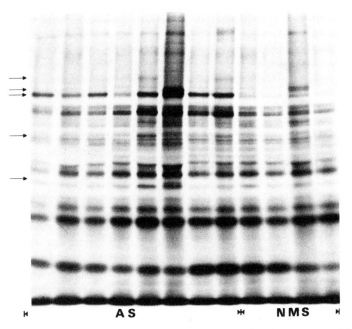

Fig. 8. Immunoprecipitation of radiolabelled adherent cell extracts
with different antisera from syngeneic mice. Adherent cells were
labelled with ^{35}S-methionine. Cellular extracts were immunoprecip-
itated by eight antisera (AS) derived from different mice immunized
with adherent cells and challenged with suspension growing cells.
Immunoprecipitation with normal mouse serum (NMS) from four different
mice served as controls. Arrows indicate antigens which either qualitatively or quantitatively distinguish the experimental group (or
parts thereof) from the control group. Approximate molecular weights
of the bands specified by the arrows (from top to bottom): 110K, 90K,
88K, 60K and 44K, respectively.

be visualized. It is noteworthy that various differences between the immune sera themselves can also be recognized. It is of interest to find out whether any of these determinants play a role in the biological differences between the suspension borne tumorigenic S49 cells and their substrate adhering, non-tumorigenic, immunogenic variants.

While the findings reported in this paper are at present mostly at the phenomenological level, they nevertheless offer a potentially significant framework to probe into the in vivo and in vitro growth regulatory mechanisms in malignant lymphoid cells of both animal and human origin.

ACKNOWLEDGEMENTS

These studies were supported in part by the Herta Schnap Fund for Cancer Research, by the United States-Israel Binational Science Foundation (BSF), Jerusalem, Israel, and by the Fund for Basic Research administered by the Israel Academy of Sciences and Humanities.

REFERENCES

1. J. Hochman, A. Katz, and Y. Weinstein, Eur. J. Cancer 15: 11 (1979).
2. J. Hochman, A. Katz, E. Levy, and S. Eshel, Cell Biol. Intern. Reports 4: 953 (1980).
3. J. Hochman, A. Katz, E. Levy, and S. Eshel, Nature 290: 248 (1981).
4. J. Hochman, N. Ben-Ishay, and M. Castel, Exp. Cell Res. 142: 191 (1982).
5. K. Horibata, and A. W. Harris, Exp. Cell Res. 60: 61 (1970).
6. P. Coffino, H. R. Bourne, J. Hochman, P. Insel, K. L. Melmon, and G. M. Tomkins, Rec. Prog. in Hormone Res. 32: 669 (1976).
7. P. A. Insel, H. R. Bourne, P. Coffino, and G. M. Tomkins, Science 190: 896 (1975).
8. V. Daniel, G. Litwack, and G. M. Tomkins, Proc. Natn. Acad. Sci. U.S.A. 70: 76 (1973).
9. P. Coffino, H. R. Bourne, and G. M. Tomkins, J. Cell. Physiol. 85: 603 (1974).
10. H. R. Bourne, P. Coffino, and G. M. Tomkins, J. Cell. Physiol. 85: 611 (1974).
11. J. Hochman, P. A. Insel, H. R. Bourne, P. Coffino, and G.M. Tomkins, Proc. Natn. Acad. Sci. U.S.A. 72: 5051 (1975).
12. I. Lemaire, and P. Coffino, Cell 11: 149 (1977).
13. R. A. Steinberg, P. M. O'Farrell, U. Friedrich, and P. Coffino, Cell 10: 381 (1977).
14. P. Coffino, J. M. Gray, and G. M. Tomkins, Proc. Natn. Acad. Sci. U.S.A. 72: 878 (1975).

15. D. K. Granner, Biochem. Biophys. Res. Commun. 46: 1576 (1972).
16. W. E. Criss, Oncology 30: 43 (1974).
17. V. S. Cho-Chung, and T. Clair, Biochem. Biophys. Res. Commun. 64: 768 (1975).
18. K. N. Prasad and P. K. Sinha, Differentiation 6: 59 (1976).
19. Y. S. Cho-Chung, and T. Clair, Nature 265: 452 (1977).
20. P. Coffino, R. Baumal, R. Laskov, and M. Scharff, J. Cell. Physiol. 79: 429 (1972).
21. R. J. Ludford, Proc. R. Soc. B 112: 250 (1932).
22. E. V. Cowdry, Archs. Path. 30: 1245 (1940).
23. D. R. Coman, Cancer Res. 4: 625 (1944).
24. F. Grinnel, Int. Rev. Cytol. 53: 65 (1978).
25. J. M. Pouyssegur, and I. Pastan, Proc. Natn. Acad. Sci. U.S.A. 73: 544 (1976).
26. J. L. Winkelhake, and G. L. Nicolson, J. Natn. Cancer Inst. 56: 285 (1976).
27. J. Bubenik, P. Perlman, E. M. Fenyo, T. Jandcova, E. Suhajova, and M. Malkowsky, Int. J. Cancer 23: 392 (1979).
28. J. C. Murray, L. Liotta, S.I. Rennard, and G. R. Martin, Cancer Res. 40: 347 (1980).
29. M. Fogel, P. Altevogt, and V. Schirrmacher, J. Exp. Med. 157: 371 (1983).
30. H. Kobayashi, ed., Gann Monograph on Cancer Research, Japan Scientific Societies, Tokyo (1979).
31. M. Shinitzky, Y. Skornick, and N. Haran-Ghera, Proc. Natn. Acad. Sci. U.S.A. 76: 5313 (1979).

TSH RECEPTOR MEDIATED GROWTH OF THYROID CELLS: MECHANISM AND

RELATION TO AUTOIMMUNE THYROID DISEASE

Salvatore M. Aloj, Claudio Marcocci, Michele De Luca,
Sidney Shifrin, Evelyn Grollman, William Valente,
and Leonard D. Kohn

Section on Biochemistry of Cell Regulation, Laboratory
of Biochemical Pharmacology, National Institute of
Arthritis, Diabetes, Digestive and Kidney Diseases,
National Institutes of Health, Bldg. 4, Rm. B1-31
Bethesda, Maryland

INTRODUCTION

Thyrotropin (TSH) is a pituitary glycoprotein hormone whose
primary role is to regulate thyroid cell function (1,2). The
interaction of TSH with a specific receptor on the thyroid cell
surface induces changes in adenylate cyclase activity which
result in the following tissue responses: enhanced iodide
uptake, thyroglobulin biosynthesis, iodination of thyroglobulin,
degradation of iodinated thyroglobulin to form thyroid hormone,
and the release of thyroid hormone (T_3 and T_4) into the blood
stream (Figure 1). Recent studies indicate that another
important action of TSH is to stimulate the growth of thyroid
cells. Initially recognized as the ability of purified TSH to
induce thymidine uptake in primary cultures of thyroid cells (3),
it has been further established by the observations that the
growth of a clonal continuous culture of functional thyroid cells
(4) is TSH dependent. This TSH effect is blocked by anti-TSH (5)
and mimicked or blocked by monoclonal antibodies directed at the
TSH receptor [as established by competitive kinetics in binding
and functional assays (6)]. Several questions emerge. For
example, is this simply another cAMP modulated response via the
same receptor structure (Fig 2A)? Alternatively are there
distinct message signals derived from a TSH interaction with a
single receptor (Fig 2B) or from two structurally distinct TSH
receptors (Fig 2C)? What is the molecular mechanism involved in
the growth stimulus?

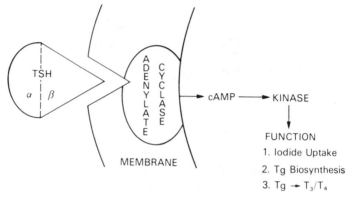

Figure 1. Classical view of TSH receptor wherein TSH interaction
with specific recognition site results in adenylate cyclase
stimulation and a sequence of 2nd message induced responses
yielding thyroid hormone. The ß subunit of TSH is depicted as
carrying the primary evidence for recognition determinants;
however evidence for subunit involvement, direct and indirect,
exists.

　　The present report will examine these questions from a
molecular point of view. First, the TSH receptor structure will
be shown to be multivalent or multicomponent using a monoclonal
antibody-reconstitution approach. Second, the same receptor
components will be shown to be involved in the growth and cAMP
signal. Third, the growth response will be shown to involve a

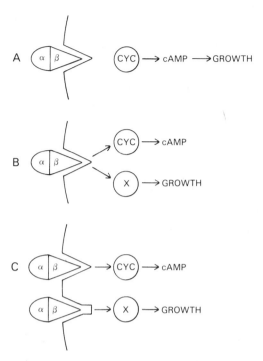

Figure 2.   Three possibilities to account for TSH receptor
mediated growth activity: (A) another cAMP directed response; (B)
independent message and response systems coupled to a single
recognition site; or (C) independent message and response systems
each with its own recognition site.

complex grouping of coregulated but independently modulated TSH
perturbations of (1) phosphoinositide metabolism involving
arachidonate as a key intermediate; (2) lipid demethylation and
cAMP modulation of membrane fluidity; and (3) alterations of
sodium-proton flux resulting in an increased intracellular pH
(Figure 3).   Though the molecular complexity of hormonally
regulated growth will be presented; provocative evidence will be
summarized that an intermediate series of reactions i.e., TSH
modulated ADP ribosylation, can also be discerned as abnormal in
growth deregulated thyroid tumors, (Figure 4).

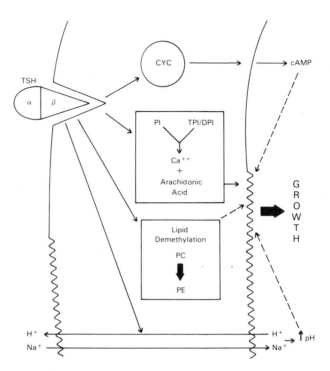

Figure 3. Signals imparted in TSH receptor mediated growth involve a series of independently coregulated biochemical events as noted. The sum does involve a change in the membrane bilayer (━━ to ⋀⋀) as well as the changes noted. The bilayer change, associated with increased fluidity, appears to be a necessary event for growth. However, the change itself does not result in growth.

Whether this reaction directly or indirectly modulates the component signals which lead to growth is not clear but will be discussed.

THE TSH RECEPTOR

Since evidence in the 1970's increasingly indicated that the recognition activity of a receptor involves molecular

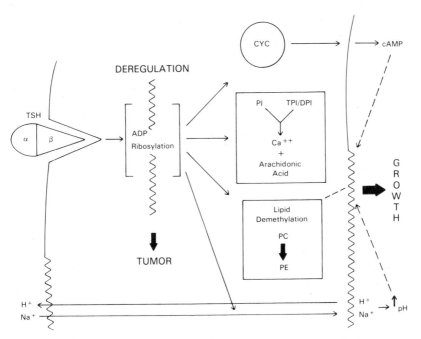

Figure 4. It is presumed that an intermediate event between
receptor recognition and the adenylate cyclase, and growth
response signals may be TSH modulated ADP ribosylation of
acceptor proteins. These act as modulators of the activities
noted in Figure 3.

elements of a membrane separate from the adenylate cyclase or
other enzyme based signalling processes (1,2), binding studies
have been used to define the TSH receptor.

The sum of the binding studies (2,7-10) identified a
membrane ganglioside, as well as a membrane glycoprotein as two
potentially important components in the cell surface recognition
event. Studies of the in vitro properties of ^{125}I-TSH binding to
each led to the following ideas concerning the process of
recognition and the mechanism by which the hormone-receptor
interaction induced a functional response in the cell. TSH
binding to the glycoprotein component of the membrane was

proposed to be the initial high affinity recognition event on the
cell surface, i.e., the necessary first step in receptor
recognition; however, a full functional response was postulated
to require the ganglioside.

The ganglioside was suggested to contribute to the
following receptor functions. It completed specificity by
distinguishing among glycoprotein hormones and related ligands
such as tetanus toxin, cholera toxin, and interferon. It
modulated the apparent affinity and capacity of the glycoprotein
receptor component and induced a conformational change in the
hormone believed necessary for subsequent message transmission.
It allowed the ligand to perturb the phospholipid bilayer through
alterations in lipid order and contributed to the ability of TSH
to alter the ion flux across the membrane. Finally, the model
proposed that after the glycoprotein receptor component trapped
the TSH much as a sperm on the surface of the ovum, the
ganglioside acted as an emulsifying agent to allow the hormone to
interact with other membrane components within the hydrophobic
environment of the lipid bilayer and thereby initiate the signal
processes. The physical basis of the emulsification process was
the formation of the anhydrous complex between the hormone and
the oligosaccharide moiety of the ganglioside. By excluding
water from the interface of the ligand-receptor complex, membrane
penetration was facilitated, and interactions involving membrane
components of the adenylate cyclase ensued. The evolution of
these ideas can be traced in a series of reviews by these authors
and their colleagues (2,7-10).

In order to see if these hypothesis derived from in vitro
experiments were true for an in vivo system, a monoclonal
antibody approach was evolved (6,11-15), i.e., the aim was to
identify monoclonal antibodies to the TSH receptor by assays of
TSH receptor function; relate these to specific membrane
molecules, i.e., the ganglioside or glycoprotein receptor
components; and then further define their functional roles. The
first step avoided a preconceived model wherein purified receptor
components were injected, but rather involved the injection of
crude solubilized thyroid membrane preparations into mice
followed by spleen cell fusion with non-IgG producing mouse
myeloma cells; production of hybridomas secreting antibodies to
the intact mouse preparation; functional identification of
antibodies related to the TSH receptor structure and, finally,
characterization of the antigenic determinants of the antibodies
with particular respect to the already identified TSH binding
components.

The second step encompassed attempts to directly relate TSH receptor structure to the autoantibodies in Graves' sera since this autoimmune disease had been linked to TSH receptor function. Altered autoimmune response has been implicated in some growth abnormalities, hence its interest in this context. This approach involved the following steps: fusion of lymphocytes from patients with active Graves' disease with a non-IgG secreting mouse myeloma cell line; identification of heterohybridomas which secreted antibodies capable of stimulating thyroid function or blocking TSH binding; and characterization of the antigenic determinants of these antibodies with respect to the structural or functional components of the TSH receptor identified by the monoclonal antibodies derived from the first approach or postulated in the receptor model.

A two-stage screening procedure with thyroid membranes was utilized. In the first stage, hybridomas producing antibodies reactive with thyroid membranes were identified by using ^{125}I-labeled staphylococcal protein A or a ^{125}I-labeled rabbit anti-mouse or anti-human IgG F(ab')$_2$ fragment-specific antibody. In the second stage, the assay was repeated as a competition assay by adding 1 μm TSH during the initial incubation of membranes and hybridoma medium. Hybridoma antibodies reactive with thyroid membranes in the first screening assay, but blocked by unlabeled TSH in the second, were chosen as potential TSH receptor antibodies.

Having identified potential anti-TSH receptor producing clones, antibody preparations were evaluated in a series of assays measuring their ability (i) to competitively inhibit ^{125}I-TSH binding; (ii) to directly stimulate adenylate cyclase activity or to act as competitive antagonists or agonists of TSH-stimulated adenylate cyclase activity; (iii) to mimic the ability of TSH to release thyroid hormones (T_3/T_4) from the thyroid in an in vivo mouse bioassay; and (iv) to stimulate iodide uptake by thyroid cells in culture. As a first approximation the following criteria were used to identify a monoclonal antibody to the TSH receptor. (i) The antibody had to inhibit TSH binding to thyroid membranes or, conversely, be itself prevented from binding to thyroid membranes by TSH. (ii) Inhibition had to be reasonably specific and had to be competitive as opposed to noncompetitive or uncompetitive. (iii) The antibody had to competitively inhibit TSH-stimulated functions or, conversely, had to mimic TSH activity and exhibit competitive agonism.

As will be noted, most antibodies to the TSH receptor,

defined by these criteria, can be broadly grouped as "inhibitors"
or "stimulators" based on these assays. An "inhibitor" has the
following characteristics: competitive inhibition of ^{125}I-TSH
binding; competitive inhibition of TSH-stimulated adenylate
cyclase activity; inhibition of TSH-stimulable thyroidal iodine
release or uptake; and no direct stimulation of thyroid adenylate
cyclase activity, T_3/T_4 release, or iodide uptake. A
"stimulator" (TSAb) in general had the following characteristics:
significantly weaker inhibition of ^{125}I-TSH biding than the first
group but equally potent competitive inhibition of TSH-stimulable
adenylate cyclase activity; direct TSH-like, stimulatory action
with respect to adenylate cyclase activity, as well as both
iodide uptake and T_3/T_4 release by the thyroid; and competitive
agonism when included with low concentrations of TSH in assays
measuring adenylate cyclase activity. An antibody with mixed
activity is a stimulatory antibody (TSAb) which also has the
ability to inhibit ^{125}I-TSH binding to a significant degree. The
significance of the existence of antibodies with "mixed"
properties will be discussed with respect to the organization of
receptor components and their active site determinants. The
antibodies discussed in this report, and their categorization as
above, are summarized in Table 1.

Table 1

Representative monoclonal antibodies to the TSH receptor

Clone No.	TSH receptor source	"Activity" pattern
13D11	bovine	inhibitor
11E8	bovine	inhibitor
22A6	bovine	stimulator
59C9	human	inhibitor
60F5	human	inhibitor
52A8	human	mixed
122G3[a]	human	inhibitor
129H8[a]	human	inhibitor
206H3[a]	human	stimulator
208F7[a]	human	mixed
307H6[a]	human	stimulator

[a] Heterohybridomas

Although all of the selected monoclonal antibodies (Table
1) had no anti-TSH activity but were able to inhibit ^{125}I-TSH

Table 2. Ability of antibodies to prevent ^{125}I-TSH binding to liposomes containing the glycoprotein component of TSH receptor[a] or to react with various ganglioside preparations [a]

Antibody	^{125}I-TSH bound (cpm)			Ganglioside reactivity measured as ^{125}I-Protein A or ^{125}I-anti-human $F(ab)_2$ binding to antibody ganglioside complexes in a solid phase assay (cpm)			
	Bovine Glycoprotein Component	Human Component	Rat Component	No added lipid	Human thyroid gang.	Rat thyroid gang.	Mixed bovine brain gang.
Controls							
Monoclonal control	18,200	14,800	29,200	150	160	155	155
N1 mouse IgG	17,940	14,700	29,500	128	185	125	135
N1 human IgG	18,400	14,800	31,200	149	144	160	145
Inhibitors							
13D11	4,100	3,800	5,400	141	235	175	215
11E8	2,100	400	2,050	156	206	210	204
59C9	6,200	1,200	5,900	121	195	195	198
60F5	4,100	1,050	4,200	110	210	210	200
129H8	1,400	980	2,800	101	285	256	170
122G3	6,400	4,200	9,200	146	268	235	155
Stimulators							
22A6	14,800	13,600	26,800	137	870	982	410
206H3	15,500	12,200	24,500	156	2,889	1,210	128
307H6	16,200	12,800	26,400	129	4,210	2,450	240
Mixed							
52A8	9,800	5,400	15,300	135	1,480	1,210	205
208F7	12,400	6,100	14,400	136	920	1,140	180

[a] Assays were performed as detailed in References (6,11-15)

Table 3. Ability of monoclonal antibodies to stimulate adenylate cyclase activity in human thyroid cells, to increase cAMP mediated iodide uptake in Rat FRTL-5 thyroid cells, or release T_3/T_4 into the bloodstream as measured by a mouse bioassaya[a]

Antibody added (0.1 mg/ml)	Adenylate Cyclase Activity		Iodide uptake in Rat FRTL-5 Thyroid Cells ^{125}I (cpm)		Mouse Blood ^{125}I (% of control)	
	Direct effect on Human Thyroid Cells — cAMP level, pmol/μg DNA	Effect on 2.5×10^{-10} M TSH activity	Alone	+TSH 2.5×10^{-10} M	Alone	+ TSH
Controls						
No addition	1.2 ± 0.2	8.8	680	7,700	130	480
Normal human IgG	0.8 ± 0.2	8.8	700	8,100	120	480
Normal mouse IgG	0.8 ± 0.2	8.8	820	7,900	118	520
Monoclonal control	0.8 ± 0.2	7.4	540	8,100	121	450
Inhibitors						
13D11	0.8 ± 0.2	3.8	700	2,450	112	170
11E8	0.8 ± 0.2	3.3	650	1,410	124	180
59C9	0.8 ± 0.2	4.1	730	2,200	110	200
60F5	0.8 ± 0.2	3.4	810	1,100	104	188
129H8	0.8 ± 0.2	3.9	740	1,500	123	210
122G3	0.8 ± 0.2	2.5	920	2,980	110	230
Stimulators						
22A6	3.0 ± 0.2	3.1	6,840	----	600	---
206H3	1.5 ± 0.2	2.9	8,200	----	440	---
307H6	3.4 ± 0.2	2.8	7,400	----	510	---
Mixed						
52A8	2.8 ± 0.2	4.2	7,800	----	340	---
208F7	1.8 ± 0.2	2.7	9,400	----	520	---

a/ Assays were performed as detailed in references 6 and 11 through 15.
All experimental values are $\pm$ 5%

binding to thyroid membranes, significant differences existed in their relative potencies. Antibodies which have been classified as inhibitors in Table 1 (11E8 and 13D11, bovine receptor; 59C9, and 60F5, human receptor; 129H8 or 122G3, Graves' autoimmune) were potent inhibitors (6,11-15) whether tested in preincubation, competition, or displacement assays, i.e., independent of the order of addition of TSH or antibody. Inhibition was competitive with respect to TSH whether measured using ^{125}I-TSH and unlabeled antibody or unlabeled TSH and unlabeled antibody, binding of antibody being measured with ^{125}I-labeled protein A.

The antibodies selected from the potent "inhibitory" group above were able to significantly block ^{125}I-TSH binding to human, bovine, or rat thyroid membranes and to liposomes embedded with the high affinity glycoprotein component of the thyroid membranes (Table 2). Using cultured thyroid cells, the antibodies noted as inhibitors in Table 1 were all unable to mimic TSH as direct stimulators of adenylate cyclase activity (Table 3). They were, however, able to inhibit TSH-stimulated adenylate cyclase activity (Table 3), inhibition in each case was competitive. The antibodies also inhibited TSH-stimulated radioiodine uptake in thyroid cells (Table 3) and TSH-stimulated T_3/T_4 release in a mouse bioassay used to measure in vivo TSH activity (Table 3).

Monoclonal antibodies 22A6, 206H3, and 307H6, were weak inhibitors of TSH binding by comparison to other antibodies (6,11-15) but were, potent inhibitors of TSH-stimulated adenylate cyclase activity (Table 3). Resolution of this contradiction evolved with the observation that each of these antibodies was a potent direct stimulator of human or rat thyroid cell adenylate cyclase activity (Table 3) and that their inhibitory action with respect to TSH was dependent on the TSH concentration. Thus, at low TSH levels, 22A6, 206H3, 307H6, 52A8, and 208F7 were more than additive competitive agonists, and at high TSH concentrations they were competitive antagonists. The antibodies were active in the mouse bioassay measuring thyroid hormone release (Table 3) and were able to enhance iodide uptake by thyroid cells in a manner identical to TSH (Table 3), i.e., under conditions where cAMP-dependent iodide uptake was the measured response.

The stimulating antibodies were all able to interact with ganglioside preparations of thyroid membranes (Table 2), whereas, with two exceptions, most were poorly reactive with the glycoprotein component (Table 2). The significance of the two exceptions, the 208F7 and 52A8 antibodies designated as mixed in Table 2 and which exhibit reactivity in both assays, will be discussed below.

The reactivity with the ganglioside preparations is relatively thyroid-specific when using total ganglioside extracts, i.e., when comparing bovine brain and bovine thyroid ganglioside preparations (Table 2). In addition, individual antibodies exhibit some species selectively for the thyroid ganglioside preparation, i.e., human better than bovine. Species specificity of autoimmune stimulators has been a well recognized phenomenon embodied in the definition of the LATS protector assay and the autoimmune stimulators (16,17).

The stimulating antibody reactivity with ganglioside preparations is lost if the glycolipid preparations is pretreated with neuraminidase (Table 4) and is highest in disialoganglioside fractions obtained by column chromatographic techniques. The highest reactivity is, however, evident with a single minor component of the disialoganglioside preparation. Incubation of this ganglioside with 1-8 thyroid tumor cells, which have had a TSH receptor defect expressed as low TSH stimulated adenylate cyclase responses correlated with a defect in ganglioside biosynthesis, can reconstitute the TSH stimulated cyclase

Table 4. Ability of monoclonal antibodies to react with modified or partially purified thyroid ganglioside preparations[a]

	Ganglioside Reactivity (cpm)	
Ganglioside Preparation	22A6	307H6
Human thyroid gangliosides	930	3,955
+ neuraminidase	210	410
Rat thyroid gangliosides	884	2,290
+ neuraminidase	195	290
Human neutral glycolipid	900	800
Monosialogangliosides	1,100	2,000
Disialogangliosides	8,050	10,700
Rat FRTL-5 disialogangliosides	11,400	22,600

[a] Prepared and assayed as detailed in References 6 and 11-15

response (Table 5). Reconstitution could also be effected by gangliosides from the cultured rat FRTL-5 thyroid cells which have a functional TSH receptor sensitive to the monoclonal antireceptor stimulators which interact with gangliosides. No reconstitution of TSH receptor function occurred in control reconstitution incubations containing no added 1-8 ganglioside extract; a ganglioside extract from the original 1-8 tumor (G_{M3});

Table 5. Reconstitution of TSH receptor expression in 1-8 rat thyroid tumors.

Thyroid Membrane or Cell Prep.	Basal	Adenylate cyclase Activity [a] p moles cAMP/µg DNA		
		Cholera Toxin $1 \times 10^{-9} M$	TSH $1 \times 10^{-9} M$	307H6 20 µg/ml
FRTL-5 thyroid cell	0.5	6.4	18	5.4
1-8 tumor (original)	0.8	0.4	0.8	0.7
+ FRTL-5 thyroid cell gangliosides	0.7	3.8	5.2	2.6
+ FRTL-5 thyroid cell disialogangliosides	0.6	1.1	7.8	5.4
+ NDase treated FRTL-5 gangliosides	0.7	1.2	0.9	0.8
+ NDase treated FRTL-5 disialogangliosides	0.7	0.5	1.9	1.2
+ mixed brain gangliosides	0.8	3.1	1.1	0.8
+ 1-8 tumor gangliosides	0.5	0.7	0.5	-
+ G_{M3}	0.8	0.7	0.7	0.8
+ FRTL-5 thyroid cell gangliosides followed by trypsin treatment[b]	0.5	-	0.9	2.9

[a]Assays were performed in triplicate; results are the average of at least three separate experiments. In no case did the standard deviation of any value exceed $\pm$ 10%.

[b]Trypsin treatment of cell preparations used the procedures referred to in References 7-15.

mixed brain gangliosides; or the FRTL-5 ganglioside extract
treated with a mixture of neuraminidases capable of converting
87% to 92% of the sialic acid residues from a lipid bound to free
form.

Incorporation of the gangliosides from the
disialoganglioside extract or from the FRTL-5 thyroid cell
preparations into the 1-8 tumor cells was monitored by reactivity
with the 22A6 or 307H6 stimulating monoclonal antireceptor
antibodies; these antibodies have been shown above to react with
thyroid ganglioside preparations (see above). Thus, for example,
as measured by ^{125}I-protein A, 22A6 binding to the no addition
cells, to cells incubated with G_{M3}, or to cells incubated with
neuraminidase treated 1-8 gangliosides was 170 $\pm$ 50 cpm/mg
membrane protein. The 22A6 binding to cells incubated with the
disialoganglioside preparation of the ganglioside mixture was 720
$\pm$ 70 and 610 $\pm$ 40 cpm, respectively (p values < 0.01 compared to
the controls above).

A cytochemical bioassay (CBA) has been used to compare the
activities of the antibodies and TSH when they were mixed
together. The cytochemical bioassay (CBA) for thyroid
stimulators is based upon quantification of changes in the
staining of sections of guinea pig thyroids, utilizing leucine-2-
naphtylamide as a chromogenic substrate to measure a lysosomal
enzyme activity. In the CBA, 22A6, 307H6, and 208F7 are
stimulators whose dose-response curves parallel these of a LATS-B
standard, a Graves' serum thyroid stimulating autoantibody
(TSAb). In contrast to 22A6, 307H6, and 208F7, 11E8 is itself
inactive as a stimulator in the CBA over a wide dose range; it
does, however, inhibit TSH stimulation in the CBA. In contrast
to its effect on TSH, 11E8 shows relatively low potency (>10,000-
fold lower) when inhibiting stimulation of the thyroid-
stimulating antibodies 22A6, 307H6, and LATS-B, i.e., even 10^2
dilutions of antibody have no effect. As anticipated, the
response to LATS-B and 307H6 is inhibited by antihuman IgG,
whereas that due to 22A6 is inhibited by antimouse IgG. Antibody
208F7 is inhibited by 10^2 dilutions of 11E8; differences in the
effect of 11E8 on TSH as opposed to 22A6, 307H6, 208F7, or LATS-B
are not the result of a 10,000-fold difference in binding
constants.

The ability of 11E8 as opposed to 22A6, 307H6, or 208F7 to
stimulate or block selectively indicates that these two
monoclonals are antibodies to different determinants on the TSH
receptor. Since separate experiments indicate that the 11E8
monoclonal antibody interacts predominantly with a membrane

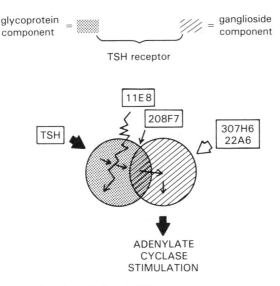

Figure 5. Hypothetical model of TSH receptor as suggested by
studies with monoclonal antibodies to the TSH receptor wherin the
receptor exists as a complex of ganglioside and glycoprotein
membrane components.

glycoprotein, whereas the 22A6 and 307H6 monoclonal antibodies
interact with a ganglioside (see above), the simplest way of
reconciling the above observations is to apply the two component
receptor model suggested in receptor binding studies (Figure 5).
Thus, TSH can be envisaged to interact first with the
glycoprotein component of the TSH receptor, which exhibits high
affinity binding properties. Its biological action, however,
requires an additional or subsequent interaction with a

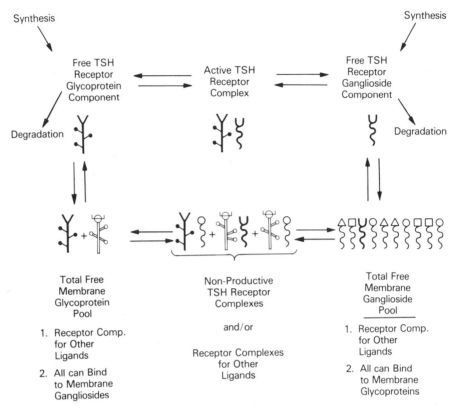

Figure 6. Hypothetical model of active TSH receptor complex in equilibrium with its free component parts which are in turn in equilibrium with other glycoproteins and gangliosides of the membrane.

ganglioside. In contrast, TSABs represented by 307H6, LATS-B, and 22A6 can be envisaged to bypass the glycoprotein receptor component and interact with the ganglioside to initiate the hormone-like signal. This model would be consistent with the observations that TSH action can be blocked by 11E8, an antibody binding to the glycoprotein receptor component. The three TSABs, 307H6, LATS-B, and 22A6, are in contrast minimally affected by 11E8 since they are directed at the next sequential step, the ganglioside, which is vital in ligand message transmission.

Antibody 208F7 raises another issue in this respect. It can interact with both ganglioside (Table 2) and glycoprotein receptor components (Table 2). It is a good inhibitor of TSH binding and a potent stimulator of adenylate cyclase activity (Table 3). It is also sensitive to 11E8 inhibition, albeit at very high concentrations by comparison to TSH. These findings suggest that there is an "overlap" or "interaction" between the two receptor components and that 208F7 interacts at this common site (Figure 5). Whether this relates to the existence of common determinants on both the glycoprotein and ganglioside or whether this relates to the fact that the "physiologic TSH receptor" is a complex of both components is not clear at this time. If, however, the receptor does exist as a complex of the two glycoconjugates, a model of a receptor complex of both components in equilibrium with free components is an intuitive extrapolation (Figure 6). Given the fact that gangliosides can be viewed as a group of similarly structured molecules with greater or lesser affinities for a particular ligand, but each with the potential for interacting with a glycoprotein − including the TSH receptor glycoprotein, an equilibrium model can be created which is far-reaching (Figure 6).

The TSH receptor has been related to receptor structures for interferon (7). Thus, each of these ligands has been associated with a receptor structure, involving gangliosides, albeit structures with different carbohydrate moieties. That interferon can influence the binding and bioactivity of TSH (7) could evolve given a receptor equilibrium model envisaged in Figure 6, i.e., perturbations of a receptor by one ligand will necessarily influence the second.

THE TSH RECEPTOR AND TSH STIMULATED GROWTH ACTIVITY

The Same Receptor Influences the cAMP Signal and the Growth Response

As noted in the introduction and as will be discussed in a separate presentation by Ambesi-Impiombato, et al, recent studies have confirmed that TSH is an important stimulus of thyroid cell growth and that this action is not directly mediated by a cAMP signal. In this respect one study (5) has shown that TSH-stimulated thymidine uptake in functioning rat thyroid cells is a valid measure of TSH-dependent thyroid cell growth; is distinct from TSH effects on cAMP levels with respect to hormone concentration, thyroid cell pretreatments, or thyroid cell species; and cannot be duplicated when dibutyryl cAMP replaces TSH. Thus TSH effects on thymidine uptake or growth appear to utilize a different receptor domain or transmission system.

The monoclonal antibodies above, as well as Graves' serum autoantibodies, have been evaluated as TSH-related "growth" promoters in thymidine uptake assays in the functioning rat thyroid cells (5,6,11-19). The data indicate that some of the "inhibitors" of the TSH-stimulated cAMP signal may be intrinsic stimulators in the growth assay in the absence of any effect on adenylate cyclase stimulation. Thus, monoclonal antibody 129H8, as well as monoclonal antibodies 22A6, 208F7, 52A8, and 206H3, were all found to be significant stimulators of $[^{14}C]$thymidine uptake. The greater role of the glycoprotein component was suggested by the fact that 129H8 was not a stimulator of adenylate cyclase (Table 3) activity nor significantly ganglioside reactive (Table 2) and the most potent stimulators, on a concentration basis, were 208F7 and 52A8, so called mixed antibodies (Table 1). Antibody 122G3 is not a significant stimulator in this experiment, but was an effective inhibitor of TSH-induced $[^{14}C]$thymidine uptake at the same concentration. This inhibitory effect has also been noted for bovine antibody 11E8 and human antibodies 59C9 and 60F5.

The ability of 122G3, 11E8, 60F5, or 59C9 to inhibit the stimulatory effect of TSH on thymidine uptake is paralleled by the inhibitory effect of each on TSH-stimulated adenylate cyclase activity. In contrast, although these four antibodies inhibit TSAb-stimulated thymidine uptake, they have no effect on TSAb-stimulated cAMP accumulation. All four antibodies are directed against the glycoprotein component of the TSH receptor. They competitively inhibit ^{125}I-TSH binding to thyroid membranes; competitively inhibit TSH-stimulated adenylate cyclase activity in thyroid cells; and have no apparent intrinsic adenylate cyclase stimulatory activity. The results obtained with the four antibodies thus suggest that the growth stimulatory effect of the TSAbs is mediated by the same TSH receptor as characterized in the studies of monoclonal antibodies using adenylate cyclase assays, i.e., Figure 2B is the best general representative schema. The data also indicate, however, that the glycoprotein receptor component may have a more functional role with respect to growth than cAMP modulation and that a different signal system must be considered.

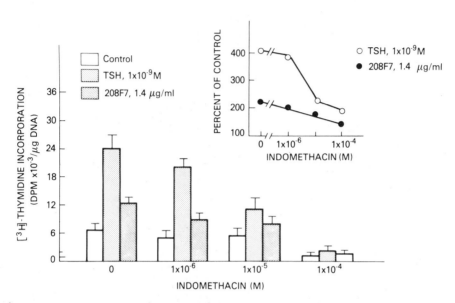

Figure 7.   Indomethacin effect on TSH monoclonal "mixed"
autoantibody (208F7) stimulated thyroid cell growth, measured by
[^{3}H]thymidine incorporation in FRTL-5 thyroid cells maintained
for 24 hours in a medium in the absence of TSH, an absolute
growth requirement.   [^{3}H]thymidine uptake was compared (i) in
cells with either no addition to the medium or with an equivalent
concentration of normal IgG (control, open bars); (ii) in cells
to whose media was added 1x10^{-9}M TSH; or (iii) in cells whose
media contained 1.4 µg/ml TSAb in addition to 0, 1x10^{-6}M, 1x10^{-4}M
indomethacin.   The inset presents the data in the body of the
figure as percent of control, i.e., as a percent of the value for
the first bar in each series of three values at 0, 1x10^{-6}M,
1x10^{-5}M indomethacin.

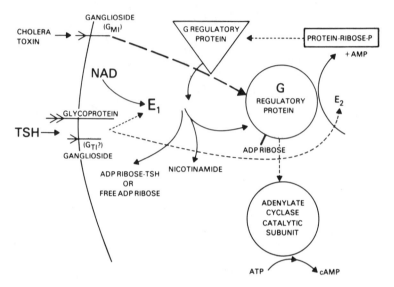

Figure 8. Proposed model wherein enzymes which regulate ADP
ribosylation of the regulatory subunit of the adenylate cyclase
are the site of hormonal control of the cAMP message transmission
process. It is suggested that TSH acts through E_1 or E_2; cholera
toxin subverts this through its own intrinsic ADP
ribosyltransferase action. It is further proposed that this also
affords a regulatory pathway whereby the cell can - with chronic
TSH stimulation, maximal G regulatory protein ribosylation, and a
stable activated state of adenylate cyclase activity - became
desensitized to further TSH stimulation and convert NAD to free
ADP ribose and nicotinamide or even ADP ribosylate TSH itself.
(See text for further discussion).

The growth response involves multiple coregulated TSH modulated signals.

The existence of a common TSH receptor together with the absence of a direct cAMP modulation of growth prompted an examination of alternate routes of signalling to affect a growth response. Several possibilities existed. First, TSH has been shown to modulate phosphatidylinositol turnover and accumulation in thyroid cells (21). A key product in PI degradation is arachidonic acid, which is in turn an intermediate in prostaglandin biosynthesis. Indomethacin, an inhibitor of arachidonic acid metabolism and prostaglandin biosynthesis blocks TSH, as well as TSAb stimulated growth as measured by thymidine uptake in thyroid cells (Figure 7), i.e., the PI path is involved.

Recent evidence indicates that PI turnover is linked to Ca^+ modulated process. Though TSH has not been shown to induce Ca^+ uptake into FRTL-5 thyroid cells, TSH modulated I^- efflux is both a Ca^+ dependent process and a process linked to PI metabolism (21). TSH can also modulate Ca^+ release from membrane preparations as measured by changes in the fluorescence of specific dyes (22). TSH induced Ca^+ release from membranes as a cell signal has also recently been linked to ligand induced changes in polyphosphoinositide metabolism in thyroid cells. Thus TSH increases phosphoinositides breakdown in these cells. Linkage of this phenomenon to growth is as yet circumstantial; however, all agents which perturb thyroid cell growth in FRTL-5 cells have thus far also been shown to perturb poly-phosphoinositide turnover.

A second TSH modulated event, important in growth is ion flux which results in altered membrane potential and internal pH. TSH modulated ion fluxes were first measured using radiolabeled triphenylmethylphosphonium (TPMP) to determine membrane potential changes (23). These were readily demonstrated as well as shown to be TSH receptor specific and cAMP independent in thyroid vesicle systems. Recent studies have linked these phenomena to thyroid cell growth; thus, amiloride, an agent which inhibits sodium-proton exchange, will block TSH and TSAb stimulated growth (Table 6). This inhibition is associated with a decrease in internal pH of the cells; a high internal pH in the thyroid cells is associated with TSH modulated growth in FRTL-5 cells.

Table 6
Effect of amiloride on TSH or TSAb stimulated growth as measured
by [^{3}H] thymidine uptake

	[^{3}H] Thymidine Uptake (cpm)	Internal pH
TSH	24,000	7.8
TSH + Amiloride	4,200	7.2
TSAb	19,400	7.9
TSAb + Amiloride	2,100	7.1

A third parameter which has been linked to cell growth is
the fluidity of the plasma membrane which is related to lipid
composition of the bilayer and which is particularly sensitive to
changes in the relative proportions of phospholipids and
cholesterol (24,25). A widely used technique to measure membrane
fluidity involves the use of the hydrophobic fluorescent dye 1,6-
diphenyl 1,3,5 hexatriene (DPH) which readily penetrates the
lipid region of plasma membranes. The polarization of
fluorescence of the dye incorporated into the membranes is
inversely related to their fluidity.

DPH polarization is low in a growing thyroid cell, i.e.,
cell growth is associated with high membrane fluidity. After TSH
withdrawal the DPH polarization rapidly increases, i.e., membrane
fluidity decreases, with cessation of growth. When TSH is
returned to the cell, there is an initial 2-3 hr duration,
further increase in polarization (or decrease in membrane
fluidity) followed by a rapid decrease in polarization (or
increase in fluidity). The second phase can be duplicated by
dibutyryl cAMP yet dibutyryl cAMP cannot itself initiate cell
growth as will be shown by Ambesi-Impiombato. Since, however,
cAMP can eliminate a prolonged 48-72 hr lag phase in TSH
stimulated growth when TSH is returned to hormone depleted cells
after 10 days, it is presumed that the cAMP effect and membrane
fluidity change is permissive, i.e., it is a necessary, although
not sufficient, membrane alteration for growth to occur. Other
non cAMP mediated, TSH dependent signals are necessary.

One can address the question of the basis for these changes. Studies using experimental conditions identical to those above show that the initial 2-3 hr TSH induced decrease in fluidity is associated with a decrease in phosphatidylethanolamine (PE) and an increase in phosphatidylcholine (PC) membrane concentrations. Altered lipid methylation has been linked to this phenomenon already by others (26). The decrease in membrane fluidity which follows TSH depletion is associated with PI changes in the membrane and an increase in cholesterol. TSH induced membrane lipid changes are thus linked to both TSH modulated cAMP and phospholipid alterations; they appear necessary but permissive in their linkage to TSH modulated growth; and they are consistent with alterations in membrane fluidity measurable with DPH fluorescence.

In sum, it is evident (Figure 4) that TSH modulated growth involves a concerted and coregulated modulation of multiple cellular processes.

Thyroid Tumors Which Result from a Loss in TSH Regulated Growth

Ideas concerning the mechanism of action of TSH with respect to the ganglioside have evolved from comparisons with cholera toxin-ganglioside interactions (2,7-10). Thus, for example, they borrow heavily on the presumptive role of the G_{M1} ganglioside to allow the α-subunit of the toxin to intercalate within the bilayer structure where its ADP ribosyltransferase activity can catalyze the ADP ribosylation of the G regulatory protein of the adenylate cyclase complex using NAD as substrate (Figure 8). Although TSH has no intrinsic ADP ribosylation activity, a series of studies (2,27,28) has, however, defined an ADP ribosyltransferase activity in thyroid membrane preparations (Figure 8). Also demonstrated in thyroid membranes was an enzyme, analogous to snake venom phosphodiesterase, which could release AMP from ADP ribosylated acceptor molecules (Figure 8). A key observation which followed was that TSH could stimulate in a dose-dependent and hormone-specific manner thyroid membrane ADP ribosyltransferase activity. Further, the result of the TSH action was to increase incorporation of [^{32}P]ADP ribose into several membrane components, including two believed to be related to the G regulatory subunit of the adenylate cyclase complex (Figure 8).

The labeling of the membrane components was further notable in several respects. First, the process appeared to precede in time the stimulation of adenylate cyclase activity. Second, the ß subunit of TSH coordinately inhibited the TSH-stimulated adenylate cyclase and ADP ribosylation activities, whereas the subunit was itself ADP ribosylated and exhibited slight cyclase stimulatory activity. Finally, conditions were uncovered wherein NAD could increase basal- and TSH-stimulated adenylate cyclase activity.

The possibility exists, however, that TSH stimulated ADP ribosylation is also linked to TSH modulated growth. Thus studies of the TSH-stimulated adenylate cyclase response have noted that after an initial exposure to TSH and stimulation of the adenylate cyclase response, the tissue can become refractory to a subsequent exposure to hormone. This process has been termed desensitization (1); the ADP ribosylation process has been linked to the desensitization process. Thus thyroid cells whose adenylate cyclase response is desensitized by chronic TSH treatment, can be resensitized if incubated with nicotinamide or arginine-L-methyl ester, an end product and artificial substrate of the ADP ribosylation reaction respectively (29,30).

Since TSH stimulated growth is a property of cells desensitized to TSH with respect to adenylate cyclase activation, the question of a link between ADP ribosylation and growth can be raised.

Thyroid tumors can be formed by chronic stimulation of TSH. These tumors effectively lose their TSH regulated growth function. Examination of two such tumors (1-5G and 1-8) have demonstrated that they also lose their TSH regulated ADP ribosyltransferase activity and have instead a 10-fold elevated "basal" ADP ribosylating activity. The possibility thus exists that deregulated ADP ribosylation can result in increased growth, i.e., that this reaction is interposed between some or all of the TSH modulated signal processes involved in growth (Figure 4).

SUMMARY

TSH receptor function has been linked to growth as well as adenylate cyclase stimulation. The same receptor structure is involved but it generates a distinct and complex set of coregulated signals. The complex linkage of TSH receptor structure to these signals has become a means for further exploring differences between differentiation and growth.

REFERENCES

1.   J. Robbins, J. E. Rall, and P. Gorden, The Thyroid and
      Iodine Metabolism, in: "Metabolic Control and Disease",
      P. K. Bondy and L. E. Rosenberg, eds., W. B. Saunders
      Company, Philadelphia (1980).

2.   L. D. Kohn, S. M. Aloj, S. Shifrin, W. A. Valente, S.
      Weiss, P. Vitti, P. Laccetti, J. L. Cohen, C. M.
      Rotella, and E. F. Grollman, The thyrotropin receptor,
      in: "Receptors for Polypeptide Hormones", B. T. Posner,
      ed., Marcel Dekker, New York (in press).

3.   R. J. Winand, and L. D. Kohn, Thyrotropin effects on
      thyroid cells in culture: effects of trypsin on the
      thyrotropin receptor and on thyrotropin-mediated cyclic
      3':5'-AMP changes, J Biol Chem 250:6534 (1975).

4.   F. S. Ambesi-Impiombato, L. A. M. Parks, and H. G. Coon,
      Culture of hormone-dependent epithelial cells from rat
      thyroids, Proc Natl Acad Sci USA 77:3455 (1980).

5.   W. A. Valente, P. Vitti, L. D. Kohn, M. L. Brandi, C. M.
      Rotella, R. Toccafondi, D. Tramontano, S. M. Aloj, and
      F. S. Ambesi-Impiombato, The relationship of growth and
      adenylate cyclase activity in cultured thyroid cells:
      separate bioeffects of thyrotropin, Endocrin 112:71
      (1982).

6.   W. A. Valente, P. Vitti, Z. Yavin,E. Yavin, C. M. Rotella,
      E. F. Grollman, R. S. Toccafondi and L. D. Kohn, Graves'
      monoclonal antibodies to the thyrotropin receptor:
      stimulating and blocking antibodies derived from the
      lymphocytes of patients, Proc Natl Acad Sci USA 79:6680
      (1982).

7.   L. D. Kohn, Relationships in the structure and function of
      receptors for glycoprotein hormones, bacterial toxins,
      and interferon, in: "Receptors and Recognition", P.
      Cuatrecasas and M. F. Greaves, eds., Chapman and Hall,
      London (1978)

8.   L. D. Kohn, S. M. Aloj, F. Beguinot, P. Vitti, E. Yavin, Z.
      Yavin, P. Laccetti, E. F. Grollman and W. A. Valente,
      Molecular interactions at the cell surface: role of
      glycoconjugates and membrane lipids in receptor
      recognition processes, in: "Membrane and Genetic
      Diseases", J. Shepard, ed., Alan R. Liss, New York
      (1982).

9.   L. D. Kohn, E. Consiglio, S. M. Aloj, F. Beguinot, M. J. S.
     De Wolf, E. Yavin, Z. Yavin, M. F. Meldolesi, S.
     Shifrin, D. L. Gill, P. Vitti, G. Lee, W. A. Valente,
     and E. F. Grollman, The structure and function of the
     thyrotropin receptor: potential role of gangliosides and
     relationship to receptors for interferon and bacterial
     toxins, in: "International Cell Biology 1980-1981", A.
     G. Schweiger, ed., Lange and Springer, Berlin (1981).
10. L. D. Kohn and S. Shifrin, Receptor structure and function:
     an exploratory approach using the thyrotropin receptor
     as vehicle, in: "Horizons in Biochemistry and
     Biophysics, Hormone Receptors", L. D. Kohn, ed., J.
     Wiley and Sons, New York (1982).,
11. E. Yavin, Z. Yavin, M. D. Schneider, and L. D. Kohn,
     Monoclonal antibodies to the thyrotropin receptor:
     implications for receptor structure and the action of
     autoantibodies in Graves' disease. Proc Natl Acad Sci,
     USA 78:3180 (1981).
12. E. Yavin, Z. Yavin, M. D. Schneider, and L. D. Kohn,
     Monoclonal antibodies to the thyrotropin receptor:
     interaction with thyroid membranes on solidsurfaces and
     on thyroid cells, in: "Monoclonal Antibodies in
     Endocrine Research," R. E. Fellows and G. Eisenbarth,
     eds., Raven Press, New York (1981).
13. L. D. Kohn, W. A. Valente, P. Laccetti, C. Marcocci, M.
     DeLuca, P. A. Ealey, N. J. Marshall and E. F. Grollman,
     Monoclonal antibodies as probes of thyrotropin receptor
     structure, in: "Receptor Biochemistry and Methodology
     - Monoclonal and antiidiotypic antibodies: Probes for
     receptor structure and function," J. C. Venter, C. M.
     Fraser and J. M. Lindstrom, eds., Alan R. Liss, New York
     (in press).
14. L. D. Kohn, E. Yavin, Z. Yavin, P. Laccetti, P. Vitti, E.
     F. Grollman, and W. A. Valente, Autoimmune thyroid
     disease studied with monoclonal antibodies to the
     thyrotropin receptor, in: "Monoclonal antibodies: Probes
     for study of autoimmunity and immunodeficiency," G.
     Eisenbarth and R. Haynes, eds., Academic Press, New York
     (in press).

15. L. D. Kohn, D. Tombaccini, M. L. DeLuca, M. Bifulco, E. F.
     Grollman, W. A. Valente, Monclonal antibodies and the
     thyrotropin receptor, in: "Receptors and Recognition:
     Antibodies to Receptors" M. F. Greaves, ed., Chapman and
     Hall, Ltd., London, England (in press).

16. D. D. Adams, Thyroid-stimulating autoantibodies, in:
     "Vitamins and Hormones," E. Diczfalusy and R. E. Olson,
     eds., Academic Press, New York (1981).

17. A. Pinchera, G. F. Fenzi, E. Macchia, P. Vitti, F. Monzani
     and L. D. Kohn, Immunoglobulins thyreostimulantes
     antigenes correspondants, Annales d' Endocrinologie
     (Paris), 520 (1982).

18. P. Laccetti, E. F. Grollman, S. M. Aloj, and L. D. Kohn,
     Ganglioside dependent return of TSH receptor function in
     a rat thyroid tumor with a receptor defect, Biochem
     Biophys Res Commun. 110:772 (1983).

19. R. H. Michell, Inositol phospholipids and cell surface
     receptor function, Biochem Biophys Acta. 415:81 (1975).

20. W. A. Valente, P. Vitti, C. M. Rotella, S. M. Aloj, F. S.
     Ambesi-Impiombato, and L. D. Kohn, Growth-promoting
     antibodies in autoimmune thyroid disease: a population
     of thyroid-stimulating antibodies measurable using rat
     FRTL-5 cells, New Eng J Med., 1983, in press.

21. S. M. Aloj, Membrane lipids and modulation of hormone
     receptor expression, in: "Horizons in Biochemistry and
     Biophysics, Vol. 6, Hormone Receptors," L. D. Kohn, ed.,
     John Wiley and Sons, New York (1982).

22. E. F. Grollman, Receptor-mediated alteration in membrane
     potential and internal pH as intracellular signals, in:
     "Horizons in Biochemistry and Biophysics, Vol. 6,
     Hormone Receptors," L. D. Kohn, ed., John Wiley and
     Sons, New York (1982).

23. E. F. Grollman, G. Lee, F. S. Ambesi-Impiombato, M. F.
     Meldolesi, S. M. Aloj, H. G. Coon, H. R. Kaback, and L.
     D. Kohn, Effects of Thyrotropin on the Thyroid Cell
     Membrane: Hyperpolarization induced by Hormone-Receptor
     Interaction, Proc. Natl. Acad. Sci U.S. 74:2352 (1977).

24. L. D. Kohn, F. Beguinot, P. Vitti, P. Laccetti, P., W. A.
     Valente, E. F. Grollman, C. M. Rotella, and S. M. Aloj,
     Thyrotropin receptor-mediated signals for differentiated
     function versus growth: implications for unregulated
     tumor growth, in: "Membranes in Tumour Growth", T.
     Galeotti, A. Cittadini, G. Neri and S. Papa, eds.,
     Elsevier Biomedical Press, Amsterdam (1982).

25. F. Beguinot, S. Formisano, C. M. Rotella, L. D. Kohn, and
     S. M. Aloj, Structural changes caused by thyrotropin in
     thyroid cells and in liposomes containing reconstituted
     thyrotropin receptor, <u>Biochem Biophys Res Commun</u>. 110:48
     (1983).

26. F. Hirata, and J. Axelrod, Enzymatic methylation of
     phosphatidylethanolamine increases erythrocyte
     membrane fluidity, <u>Nature</u>. 275:219 (1978).

27. M. De Wolf, P. Vitti, F. S. Ambesi-Impiombato, and L. D.
     Kohn, Thyroid membrane ADP-ribosyltransferase activity:
     stimulation and activity in functioning and non-
     functioning thyroid cells, <u>J Biol Chem</u>. 256:12287
     (1981).

28. P. Vitti, M. J. S. DeWolf, A. M. Acquaviva, M. Epstein, and
     L. D. Kohn, Thyrotropin stimulation of the ADP
     Ribosylation Activity of bovine thyroid membranes, <u>Proc
     Natl Acad Sci USA</u>. 79:1525 (1982).

29. S. Filetti, N. A. Takai, and B. Rapoport, Prevention by
     Nicotinamide of desensitization to thyrotropin
     stimulation in cultured human thyroid cells, <u>J. Biol.
     Chem.</u> 256:1072 (1981).

30. S. Filetti, and B. Rapoport, Hormonal Stimulation of
     Eucaryotic Cell ADP Ribosylation: Effect of Thyrotropin
     on Thyroid Cells, <u>J. Clin. Invest.</u> 68:461 (1981).

THE MITOGENIC ACTIVITY OF THYROTROPIN: THE ROLE OF cAMP

D. Tramontano, R. Picone, B.M. Veneziani, G. Villone and
F.S. Ambesi-Impiombato

Centro di Endocrinologia ed Oncologia Sperimentale del
C.N.R., c/o Dipartimento di Biologia e Patologia
Molecolare, II Policlinico, Via S. Pansini, 5 - 80131
Napoli (Italy)

INTRODUCTION

The very existence and the evolution of multicellular organisms has been made possible by continuous and finely regulated
coordination between the various cell types which constitute their
tissues, organs and systems. In complex organisms, some of the
different kinds of cells are connected physically and biochemically in very direct ways (cell-cell contacts of a very speciali-
zed kind as found in the central nervous system, neuromuscular
junctions etc.). However, most of the coordination between the
different cells which are located far apart in the various
districts of the body depend on a variety of specific chemical
messages. What may be considered the prototype of intercellular
comunications in the phylogenesis takes place within the colonies
of a unicellular organism, the slime mold Dictyostelium Discoideum. The various cells within a colony differentiate as a consequence of cell contacts and aggregation. Interestingly, the chemical signal for this cell aggregation, the prototype of a hormone,
happens to be c-AMP (Loomis, 1982).

Clinical and experimental, animal studies in vivo have long
ago revealed the existence of now classical hormones which are
secreted in the blood by the various, well described endocrine
glands. These, by complex feedback mechanisms, finely regulate the
activity of specific target organs, and thus, the various body
functions. Lately, however, classical Endocrinology seemed almost

197

to be entering a golden age in which most of the complex inter-
connections between different hormones and endocrine glands have
been described, confirmed and well established both under physiological and pathological conditions in various animals (Experimental Endocrinology) and in man (Clinical' Endocrinology). At the
same time, however, a newer branch of science, Cell Biology, with
its most recent advances, is entering the field (Fig. 1). As a
result, chemical and hormonal interactions can be studied on individual cells in vitro (Cellular Endocrinology). On in vitro model
systems of cultured eukaryotic cells, a new array of biochemical
messengers, regulatory and growth factors have been and are being
described in the literature.

Concepts such as classical hormones and growth factors need to
be modified and the borderline between the two is rapidly fading.
In evolution, recent reports (Le Roith D. et al., 1890) show us
that hormones themselves are very old, as they greatly preceded
the appearance of endocrine cells, target organs, and even the
appearance of differentiated cells and multicellular organisms. E.
Coli and other non-eukaryotes were using molecules such as insulin
to be used much later as a hormone for obviously different rea-
sons. As a consequence, the concept of hormones should not, in our
opinion, be automatically associated with physically confined
endocrine glands and target organs (the latter never existed for
classical hormones such as the thyroid hormones T_3 and T_4).

THE CELL BIOLOGY APPROACH

Several difficulties - the reported limited growth in vitro of
normal cells (Hayflick L., 1965), the unavailability, up to recent
times, of normal and differentiated cell strains in vitro, and the
ubiquitous presence of fetal and adult calf serum in large amounts
(20-30%) in cell culture media greatly prevented the use of cell
cultures in this field.

In vivo studies however were greatly limited due to the fact
that surgical deprivation of several important endocrine glands at
the same time was unfeasible in animals, and even ablation of one
important gland obviously led to uncontrollable side effects when
a specific hormone or target tissue was under study.

Newly formulated, rich culture media (McKeehan W. et al.,
1976), substitution of animal serum with hormones and growth factors in hormonally defined, serum-free media (Hayashi I. and Sato
G.H., 1976), rendered the cell culture approach both feasible and
necessary. Some recently discovered factors (NGF, FGF, EGF etc.)
are already classical. With this approach, a) new substances, like
those just mentioned, have been discovered to have hormonal-like
actions and b) new functions are being discovered for well-known
and defined hormones. For other hormones, and Thyrotropin (TSH) is

among them, in vitro systems have been and are presently being used to further clarify their biological role and their intracellular mode of action.

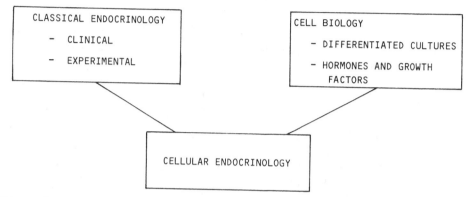

Fig. 1.   The merging of classical endocrinology and cell biology, with its latest acquisitions, makes possible the study of pure populations of endocrine cells - hormone producers or target organs, or both - in vitro.

TSH ACTIONS

TSH is known to exert its specific stimulatory action, in animals and in man, on thyroid follicular cells. Produced and secreted by Beta-2 cells of the anterior pituitary, its release is finely controlled by complex feedback regulations involving thyroid hormones (T_3 and T_4) levels and the specific hypotalamic Thyrotropin Releasing Factor (TRF).

Hormonal mode of action, in the case of TSH as well as in the case of other hormones, has not been elucidated at subcellular and molecular levels, but again, as in the case of other hormones, the role of cyclic AMP (cAMP) as intracellular "second messenger" is widely recognized. TSH specificity towards its target cells is considered not due to the rather ubiquitous TSH receptor on cell plasma membrane but rather to the thyroid-specific TSH-activated

adenylate-cyclase (Pastan I. and Macchia V., 1967; Moore W. and
Wolff I., 1974).

Among the various effects of TSH on thyroid cells, all finali-
zed to the control of the thyroid functions and ultimately leading
to the production and secretion of T_3 and T_4, the most
important are:

1) TSH receptor clustering and internalization, demostrated in
vitro by Tramontano et al. (Avivi A. et al., 1982);

2) The already mentioned stimulation of the thyroid adenylate-
-cyclase system, demonstrated in vivo (Pastan I. and Macchia V.,
1967) and in vitro (Mandato E. et al., 1981);

3) Several cAMP-mediated actions, such as:
   a) stimulation of I^- uptake and metabolism (Tong W.,
      1974);
   b) production, iodination and release of Thyroglobulin
      (Tg) in the follicular lumen;
   c) Tg uptake from the follicular lumen by simple pinocy-
      tosis or receptor-mediated endocytosis (Shifrin S. et
      al., 1982), and Tg breakdown;
   d) thyroid hormones secretion.

TSH AS A THYROID-SPECIFIC MITOGEN

An additional effect of TSH - stimulation of thyroid follicular cell division - has been postulated but in vivo experimental
systems and clinical studies did not give a definitive answer
(Westermark B. et al., 1979; Nitsch L. and Wollman S.H., 1980).

Table I

PARAMETER	CELL LINE	
	FRTL	FRTL-5
TUMORIGENICITY	-	-
CONTACT INIBITION	+	+
GROWTH IN AGAR	-	-
I^- TRAPPING	+	+
Tg PRODUCTION	+	+
TSH REQUIRED FOR GROWTH	+	+
SERUM CONCENTRATION IN MEDIUM	0.5 %	5 %
POPULATION DOUBLING TIME	4-6 Days	31 Hrs

The availability of differentiated and normal cell lines derived from rat thyroid follicular cells permanently growing in culture, has afforded a new insight with respect to the mitogenic role of TSH.

These cell lines are the FRTL (Ambesi-Impiombato F.S. et al., 1980) and the fast-growing variant FRTL-5 (Ambesi-Impiombato F.S. et al. 1982). Apart from the growth rate, they are similar to each other (Table I) in all differentiative and normal parameters and they maintain, after years of continuous culture, an absolute requirement for TSH.

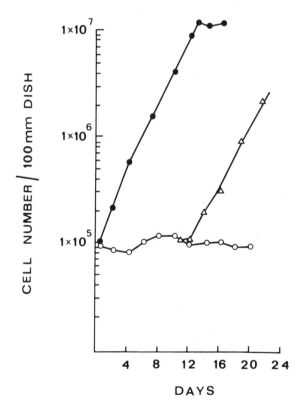

Fig. 2.   Growth curves of FRTL-5 cells in 6H, full hormonal
            complement (●-●), in 5H medium, lacking TSH (O-O)
            and after readdition of TSH at day 10 (△-△).

FRTL-5, as well as FRTL cells, are kept in hormonally supplemented medium (6H medium) among which TSH is present in fairly
elevated amounts (presently 1 mU/ml). Withdrawal of TSH from medium results in drastic changes of cell shape and morphology, as a
consequence of extensive rearrangements of cytoskeletron elements,
selectively in the microfilament (actin) component (Tramontano et
al. 1982). The effect on cell proliferation is equally dramatic:
cell growth, with a population doubling time of 31 hrs in the continuous presence of TSH, is completely arrested upon withdrawal of
the hormone (cells kept in 5H medium, lacking TSH). Growth is resumed at a normal rate after TSH readdition (Fig. 2).

The latter is equally effective after several-weeks-long starvation periods. This proves that cells, albeit non-proliferating
and quiescent, are still fully viable in the absence of TSH.

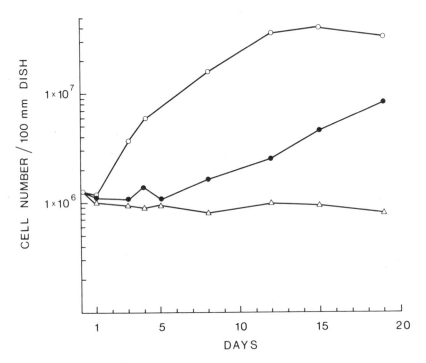

Fig. 3.   Growth curves of FRTL-5 cells in control medium (6H)
           (O-O), in 5H (lacking TSH) medium (Δ-Δ), and 5H
           medium in the presence of 10^{-3} M $(Bu)_2$ cAMP (●-●).

THE ROLE OF CYCLIC AMP

TSH receptor and cyclase activities are still present during
TSH starvation, and furthermore sensitivity to the hormone increases under these circumstances to a plateau during the first 10

days. This renders the FRTL-5 cells a clinically useful and standardized tool in the bioassay of TSH and of TSH analogues, such as anti-thyroid antibodies present in the serum of Graves and of autoimmune thyroid diseases (Vitti P. et al., 1982, and Aloj S.M. et al., in this volume).

The existence itself of proliferating FRTL/FRTL-5 cells in culture, and the wide range of data, very recently accumulated using this _in vitro_ system, clearly demonstrate the mitogenic activity of TSH. Furthermore, the dissociation between adenylate--cyclase stimulation (present) and the growth stimulation (absent) in TSH-starved FRTL-5 cells demonstrate the existence of separate hormonal bioeffects (Valente et al. 1983).

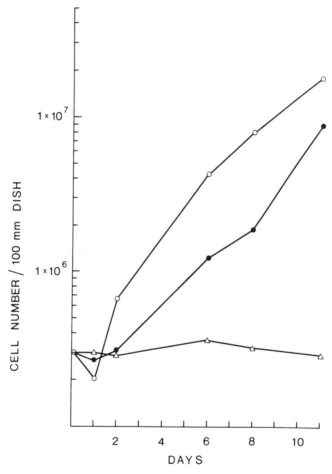

Fig. 4. Growth curves of FRTL-5 cells in control medium (6H) (O-O), in 5H medium (Δ-Δ) and in 5H medium in the presence of 10^{-3} M 8-Br-cAMP. (●-●)

Consequently, not all the effects of Thyrotropin, at the level of the target cells, seem to be mediated by cAMP.

Substitution of TSH by the cAMP analogue Dibutyryl-cAMP $((Bu)_2cAMP)$ failed to completely reproduce the effects of TSH on cell growth (Fig.3). Within the first 6-8 days, FRTL-5 cells do not proliferate in 5H medium (lacking TSH) and $(Bu)_2cAMP$ at concentrations between 10^{-2} and 10^{-9} M. After this lag phase, cell growth was resumed only in the presence of 10^{-3} M $(Bu)_2cAMP$. Growth rate, however, proceeds at a sensibly lower rate if compared to cells in 6H. Control experiments ruled out any effect of Na-butyrate.

Moreover, the effect of a different cAMP analogue, the 8-Br-cAMP, was tested on our cell system (Fig. 4). 8-Br-cAMP shares with $(Bu)_2cAMP$ the permeability through cell plasma membranes and is in addition reportedly more resistant to intracellular phosphodiesterases. Growth of FRTL-5 cells is stimulated by 8-Br-cAMP without a lag phase, but again at a slower rate - even at the maximally active 10^{-3} M concentration - when compared to the cells in regular 6H medium.

In conclusion, the experimental evidence described here demonstrate that the interaction of TSH with its natural target cell is particularly complex, and only by the use of in vitro cultured differentiated cells will it be possible to dissect the hormonal effects into different subcellular or molecular pathways. Differentiated systems in vitro may also be used after characterization of the hormonal control mechanisms to study the genetics of cell differentiation in somatic cells. In vitro systems also offer the possibility of creating genetic mutants in which hormonal control mechanisms and cell functions may be altered. Mutants of thyroid functions will expand the possible utilization of the normal thyroid cultured cells in the study of the hormonal control of metabolism. These obviously include both unspecialized cellular functions such as cell proliferation and cAMP metabolism, and thyroid-specific functions such iodide active transport and metabolism, thyroglobulin synthesis and thyroid hormone production.

ACKNOWLEDGEMENTS

We are grateful to Michele Mastrocinque and Cinzia Di Marino for skillful tecnical assistance. Part of this work has been supported by CNR, through Progetto Finalizzato Ingegneria Genetica.

REFERENCES

Ambesi-Impiombato F.S., Parks L.A.M. and Coon H.G., 1980. Culture
    of hormone-dependent funcional epithelial cells from rat
    thyroids. Proc. Natl. Acad. Sci. USA 77: 3455.

Ambesi-Impiombato, Picone R. and Tramontano D., 1982. Influence of
    hormones and serum on growth and differentiation of the thy-
    roid cell strain FRTL. CSH Conf. on Cell Proliferation 9: 483.

Avivi A., Tramontano D., Ambesi-Impiombato F.S. and Schlessinger
    J., 1982. Direct visualization of membrane clustering and
    endocytosis of thyrotropin into cultured thyroid cells. Molec.
    and Cell. Endocrinol., 25: 55.

Hayashi I. and Sato G.H., 1976. Replacement of serum by hormones
    permits growth of cells in a defined medium. Nature, 259: 131.

Hayflick L., 1965. The limited in vitro lifetime of human diploid
    cell strains. Exp. Cell Res., 37: 614.

Le Roith D., Shiloach J., Roth J. and Lesniak M.A., 1980.
    Evolutionary origins of vertebrate hormones: Substances simi-
    lar to mammalian insulins are native to unicellular euka-
    ryotes. Proc. Natl. Acad. Sci. USA, 77: 6184.

Loomis W.F., ed. in: The development of Dictyostelium discoideum.
    Academic Press, N.Y., 1982.

Mandato E., Catapano R., Ambesi-Impiombato F.S. and Macchia V.,
    1981. Cyclic nucleotide metabolism in differentiated and
    undifferentiated epithelial thyroid cells in culture. Biochem.
    Biophis. Acta, 676: 91.

McKeehan W., Hamilton W.G. and Ham R.G., 1976. Selenium is an
    essential trace nutrient for growth of WI-38 diploid human
    fibroblasts. Proc. Natl. Acad. Sci. USA, 73: 2023.

Moore W. and Wolff I., 1974. Thyroid-stimulating hormone binding
    to beef thyroid membranes. J. Biol. Chem. 281: 6255.

Nitsch L. and Wollman S.H., 1980. Thyrotropin preparations are
    mitogenic for thyroid epithelial cells in follicles in
    suspension culture. Proc. Natl. Acad. Sci. USA, 77: 2743.

Pastan I. and Macchia V., 1967. Mechanisms of thyroid-stimulating
    hormone action. J. Biol. Chem., 242: 5757.

Shifrin S., Consiglio E., Laccetti, P., Salvatore G. and Kohn L.,

1982. Bovine thyroglobulin. 27 S iodoprotein interactions with thyroid membranes and formation of a 27 S iodoprotein in vitro. J. Biol. Chem. 257: 9539.

Tong W., 1974. The isolation and culture of thyroid cells. In: Methods in Enzimology, 32: 745.

Tramontano D., Avivi A., Ambesi-Impiombato F.S., Barak L., Geiger B. and Shlessinger J., 1982. Thyrotropin induces changes in the morphology and in the organization of microfilament structures in cultured thyroid cells. Exp. Cell Res. 137: 269.

Valente W.A., Vitti P., Kohn; L.D., Brandi M.L., Rotella C.M., Toccafondi R., Tramontano D., Aloj S.M. and Ambesi-Impiombato F.S., 1983. The relationship of growth and adenylate cyclase activity in cultured thyroid cells: separate bioeffects of thyrotropin. Endocrinology, 112: 71.

Vitti P., Valente W.A., F.S. Ambesi-Impiombato, Fenzi G.F., Pinchera A. and Kohn L.D., 1982. Graves' IgG stimulation of continuously cultured rat thyroid cells: a sensitive and potentially useful clinical assay. J. Endocrinol. Invest. 5: 179.

Westermark B., Karlsson F.A. and Walinder O., 1979. Thyrotropin is not a growth factor for human thyroid cells in culture. Proc. Natl. Acad. Sci. USA, 76: 2022.

RELEVANCE OF CELLULAR POLYAMINES IN THE RESPONSE OF HEART CELLS TO CYCLIC NUCLEOTIDE-MEDIATED EFFECTORS

C. Clo, B. Tantini, C. Pignatti, S. Marmiroli, and C.M. Caldarera

Institute of Biochemistry, School of Medicine and Surgery, University of Bologna

Cyclic AMP (cAMP) and cyclic GMP (cGMP) are recognized as intracellular mediators of many hormones and chemicals which act on the heart cell (HC). While cAMP appears to be associated with a positive inotropic effect and is mainly related to β-adrenergic receptors, cGMP has closer connections with the muscarinic receptors and seems to represent a negative signal for the rate and intensity of contraction[1]. Furthermore, in a variety of normal or neoplastic cell types, the two nucleotides have been shown to be important regulators of cell growth and differentiation, with cGMP representing a positive signal for the induction of cell proliferation and cAMP a negative signal[2]. Our recent research performed on chick embryo HC cultures have indicated that the naturally occurring polyamines (PAs) spermine (SPM), spermidine (SPD) and putrescine (PTC) play an important role in the control of cellular cyclic nucleotide contents. Exogenously administered PAs cause a decrease of cAMP and counteract the action of different cAMP-mediated effectors by increasing cAMP-phosphodiesterase activity (cAMP-PDE)[3] and by reducing adenylate cyclase activity[4]. Conversely, cGMP significantly accumulates shortly after the addition of each amine to quiescent and serum-starved HC as a consequence of an increased guanylate cyclase activity and a reduced cGMP-PDE activity[5]. On the other hand, the treatment of HC with specific inhibitors of PA biosynthesis leads to the accumulation of cAMP and to the depletion of cGMP[6]. If compared with PA-rich cells, PA-deficient cells exhibit higher adenylate cyclase and cGMP--PDE activities but significantly lower guanylate cyclase and cAMP--PDE activities[7], thus supporting a role of cellular PAs in the regulation of cyclic nucleotide metabolism. The property of PAs to influence cellular cAMP and cGMP contents might constitute a basis for their positive correlation with the proliferative phenomena[8] as well as for many of their effects on development, differentiation

and function of various cell types[9]. The aim of this work was to
investigate whether the sensitivity of HC to hormones and chemicals
whose action involves early changes in cAMP and/or cGMP is also defined by the intracellular amount of PAs. In this study, primary chick
embryo HC cultures were utilized at the confluency stage in order
to reduce the possible interference of pleiotypic effects with the
specific effects under investigation.

METHODS

Experimental Procedure

Primary beating heart cells from 10 day-old chick embryos were
grown as monolayers as described elsewhere[10]. When complete confluency was reached, groups of 4 plates (for each experimental condition) were mantained for 20 hrs in the absence or presence (0.5% or
10%) of foetal calf serum and of 1 mM α-difluoromethylornithine,
kindly provided by the Centre de Recherche Merrel International
(Strasbourg, France). Cells were then fed with 4 ml of fresh medium
supplemented with 20% foetal calf serum or 1 μM morphine or 1 μM
insulin or 5 μM prostaglandin E[1].

Polyamine and Cyclic Nucleotide Analysis. At the end of the incubation periods the plates were rinsed twice with cold 0.85% NaCl,
harvested by scraping in 0.6M $HClO_4$, frozen and thawed twice and
centrifuged at 15,000 g for 10 min. Aliquots of the supernatants
were assayed for PAs and cyclic nucleotides, while the pellets were
dissolved in 1 N NaOH and assayed for protein[11]. Polyamines were
quantified by a high-pressure liquid chromatographic technique[12].
The total polyamine content was calculated as (2 x PTC content) +
(3 x SPD content) + (4 x SPM content) and is expressed as nanoequi-
valents of N/mg of protein (neq N/mg protein). Cellular cAMP and
cGMP were assayed in duplicate by using the kits from Radiochemical
Centre, after purifying the cyclic nucleotides as previously reported[5].

RESULTS AND DISCUSSION

The recent development of specific inhibitors of Pa synthesis
has greatly contributed in evaluating the role of these polycations
in a variety of cellular processes. Our recent results obtained from
HC cultures by using α-difluoromethylornithine (DFMO), an irrevers-
ible inhibitor of ornithine decarboxylase, the initial and rate-limiting enzyme in PA biosynthetic pathway, have indicated that the
steady-state content of cyclic nucleotides as well as the activity
of their metabolic enzymes might be related to the amount of endogenous PAs[6,7]. To understand whether endogenous PAs are also relevant
in defining the responsiveness of HC to different cyclic nucleotide-
mediated effectors, cellular populations containing different
amounts of PAs were utilized. For this, confluent HC cultures were

treated for 20 hrs with increasing concentrations of serum (0%, 0.5% or 10%), which is known to stimulate PA synthesis[5,10], in the absence or presence of 1 mM DFMO. The data reported in Table 1 show that with respect to serum-starved cells, increasing doses of serum cause progressive accumulation of PTC, SPD and SPM. This is paralleled by a progressive reduction of the cAMP/cGMP ratio due to the concomitant decrease of cAMP and increase of cGMP contents. On the other hand DFMO-treated cells exhibit lower levels of PAs and significantly higher cAMP/cGMP ratios than their respective counterparts treated with serum alone, thus suggesting that PA accumulation is a requirement for the action of serum on cyclic nucleotides. A similar relationship between PA, cAMP and cGMP contents has been observed by us in phospholipid-stimulated HC[13].

Table 1. Effect of Serum and 1mM DFMO on Polyamine and Cyclic Nucleotide Contents of Heart Cell Cultures

Addition (20 hrs)	PTC	SPD	SPM	PTC+SPD+SPM	cAMP	cGMP	$\dfrac{\text{cAMP}}{\text{cGMP}}$
	(nmol/mg protein)			(neqN/mg prot)	(pmol/mg prot)		
No serum	3.7	6.2	9.1	62.4	24.20	1.09	22.20
0.5% serum	4.4	8.0	10.7	75.6	20.00	1.22	16.39
0.5% serum + DFMO	2.0	4.5	9.0	53.5	26.20	0.94	27.87
10% serum	6.1	12.4	14.0	105.4	16.78	1.96	8.56
10% serum + DFMO	4.2	5.8	8.9	61.4	22.02	1.02	21.58

Experimental conditions were described in Methods. Data are the means of six determinations from 3 separate samples, differing by no more than 10%.

The intracellular contents of cAMP and cGMP of cultured cells are rapidly modified by the presence of several agents, including serum, insulin, morphine and PGE_1 in the incubation medium[14]. Table 2 shows that after a 20 hr preincubation with or without DFMO a decrease in cAMP and an increase in cGMP content occurs by shifting the cells to fresh medium containing 20% serum, resulting in a fall of the cAMP/cGMP ratio. However the magnitude of this fall is directly related to the amounts of PAs present in the cells (see Table 1). Compared with cells pretreated with serum alone, cells pretreated also with DFMO exhibit a much lower capacity to respond to 20% serum addition.

Cyclic nucleotides are involved in the action of opiate narcotics, such as morphine. A decrease in cAMP content and an increase in cGMP has been observed in many systems after its addition[14,15]. The data reported in Table 3 indicate that similar changes in cAMP and cGMP also occur in morphine-treated HC cultures. It can be seen that while in serum-starved cells the cAMP/cGMP ratio falls more

Table 2. Effect of 20% Foetal Calf Serum on Cyclic Nucleotide Contents of Heart Cells Pretreated With or Without Serum and 1 mM DFMO

Pretreatment (20 hrs)	20% Serum Addition (min)	cAMP	cGMP	$\frac{cAMP}{cGMP}$	%
		(pmol/mg protein)			
No Serum	0	24.20	1.09	22.20	100
	10	16.51	2.01	8.21	36.99
0.5% Serum	0	20.00	1.22	16.39	100
	10	13.50	2.62	5.15	31.43
0.5% Serum + DFMO	0	26.20	0.94	27.87	100
	10	22.30	1.40	16.28	58.41
10% Serum	0	16.78	1.96	8.56	100
	10	7.97	5.14	1.55	18.10
10% Serum + DFMO	0	22.02	1.02	21.58	100
	10	18.07	1.75	10.33	47.86

After a 20 hr preincubation with or without serum and DFMO the cells were fed with fresh medium supplemented with 20% foetal calf serum. Data are the means of six determinations from 3 samples, differing by no more than 10%.

Table 3. Effect of 1 μM Morphine on Cyclic Nucleotide Contents of Heart Cells Pretreated With or Without Serum and 1 mM DFMO

Pretreatment (20 hrs)	Morphine (min)	cAMP	cGMP	$\frac{cAMP}{cGMP}$	%
		(pmol/mg protein)			
No Serum	0	24.20	1.09	22.20	100
	15	19.74	1.52	12.98	58.49
0.5% Serum	0	20.00	1.22	16.39	100
	15	14.97	2.86	5.19	31.66
0.5% Serum + DFMO	0	26.20	0.94	27.87	100
	15	22.64	1.34	17.15	61.54

Experimental conditions were as in Table 2.

than 41%, the fall is more than 68% in 0.5% serum-stimulated cells containing higher PA levels. Pretreatment with DFMO while preventing PA accumulation by 0.5% serum (see Table 1) significantly reduces the capacity of morphine to decrease cAMP and to increase cGMP contents.

The possibility that the actual amount of cellular PAs may contribute in defining the sensitivity of HC to cyclic nucleotide-mediated effectors is further supported by the data obtained by using insulin (Table 4). Both in serum-starved cells and in cells pretreated with 10% serum + DFMO the addition of 1 μM insulin promotes a decrease of cAMP and an increase of cGMP content. Again the ability of the hormone to modulate the two nucleotides is greatest in cells containing the highest amount of PAs (10% serum-pretreated cells) and

Table 4. Effect of 1 μM Insulin on Cyclic Nucleotide Contents of
Heart Cells Pretreated With or Without Serum and 1 mM DFMO

Pretreatment (20 hrs)	Insulin (min)	cAMP	cGMP	$\dfrac{\text{cAMP}}{\text{cGMP}}$	%
		(pmol/mg protein)			
No Serum	0	24.20	1.09	22.20	100
	10	16.73	1.25	13.38	60.30
10% Serum	0	16.78	1.96	8.56	100
	10	8.65	3.12	2.77	32.38
10% Serum + DFMO	0	22.02	1.02	21.58	100
	10	19.69	1.30	15.10	70.02

Experimental conditions were as in Table 2.

is strongly reduced after preincubation with the PA inhibitor.

    Contrary to serum, morphine and insulin, PGE_1 increases cellu-
lar cAMP levels by stimulating adenylate cyclase activity in a number
of cell types[1,14,16].

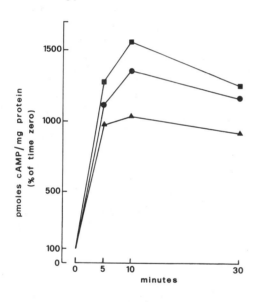

Fig.1. Effect of 5 μM PGE_1
on cAMP content of heart
cells pretreated with or
without serum and 1 mM DFMO.
Experimental conditions
were as in Table 2.
The cAMP contents at zero
time are shown in Table 1.
Preincubations: (●) no
serum; (▲) 10% serum;
(■) 10% serum + DFMO.

Figure 1 shows that a large increase of cAMP content of HC occurs
shortly after the addition of 5 μM PGE_1, with a maximum within 10 min.
No significant changes of cGMP content were observed under the same
experimental conditions. It can be seen that with respect to serum-
starved cells, pretreatment with 10% serum, while increasing PA
content (see Table 1) significantly reduces the responsiveness of
HC to PGE_1. On the other hand, DFMO, while preventing PA accumulation by serum, enhances the cellular sensitivity to PGE_1 with respect
to 10% serum alone-treated and to serum-untreated cells.

A similar increased sensitivity of PA-deficient HC toward a cAMP-
-mediated effector, such as noradrenaline, has been observed by us[10]
after pretreatment with horse serum in the presence or absence of
α-methylornithine, a reversible inhibitor of ODC and methylglyoxal
bis (guanylhydrazone) which prevents SPD and SPM accumulation. Preliminary data also indicate that by pretreating the cells with PTC,
SPD or SPM at concentrations in the range of those found in physiological fluids, the cells fail to increase cAMP content after addition of noradrenaline, further suggesting that the increase of cellular PA pool negatively influences the ability of HC to respond to
the cAMP-mediated hormone.

CONCLUSIONS

     Our data indicate that in HC cultures the accumulation of PAs
is associated with an increased capacity of the cells to respond to
agents (serum, insulin and morphine) which cause the fall of cAMP
content and the increase of cGMP. Conversely, PA-rich cells strongly
reduce their responsiveness to cAMP-mediated effectors, such as PGE[1]
and noradrenaline. The differences observed between PA-rich and PA-
-deficient cells in response to cyclic-nucleotide mediated effectors
are possibly related to the differences in the steady-state activity
of the specific enzymes, as cyclases and phosphodiesterases,of cyclic
nucleotide metabolism, observed in these cells by us[6,7]. At the present
it is only possible to speculate as to the significance of these
findings. High concentrations of PAs are characteristic of rapidly
growing systems[8]. Tumor cells or cells transformed by oncogenic viruses or by carcinogenetic compounds contain higher amounts of PAs
than normal cells[8,17]. Transformation is also associated with decreased cAMP and increased cGMP levels in many cell types[16,18], as
the result of alterations in their specific metabolic enzymes. Furthermore in transformed cell cultures as well as in proliferative
diseases, including cancer, there is a generalized decreased sensitivity to agents which elevate cellular cAMP level[18,19,20,21,22],
such as cathecolamines, isoproterenol, PGE[1], glucagon, ACTH, TSH or
MSH, probably as the result of alterations in the structure and composition of the plasma membrane. Conversely, transformed or malignant
cells escape from normal growth control by increasing their responsiveness to growth promoting agents[17,23], such as serum or insulin.
On the basis of these observations we are tempted to suggest that
the changes in the activity of cyclic nucleotide enzymes as well as
the impaired or improved responsiveness of rapidly growing tissues
or cells to cyclic nucleotide-mediated effectors might also be related
to the increased rate of synthesis and accumulation of PAs usually
observed in these systems. Experiments are in progress to verify this
hypothesis.

ACKNOWLEDGMENTS

The skillful technical assistance of Mr. S. Manfroni is gratefully acknowledged. This work was supported by a grant from Consiglio Nazionale delle Ricerche, Roma, Italy.

REFERENCES

1. H. Vapaatalo, T. Metsä-Ketelä, J. Parantainen, T. Palo-Oja, M. Kangasako, and K. Laustiola, The role of cyclic nucleotides and prostaglandins in heart function, Acta Biol. Med. Germ.,37:785 (1978).

2. I.H. Pastan, G.S. Johnson, and W.B. Anderson, Role of cyclic nucleotides in growth control, Ann. Rev. Biochem.,44:491 (1975).

3. C. Clô, B. Tantini, M.N. Coccolini, and C.M. Caldarera, Involvement of polyamines in cyclic AMP metabolism in heart cell cultures, Adv. Polyamine Res.,3:333 (1981).

4. C. Clô, B. Tantini, C. Pignatti, S. Marmiroli, and C.M. Caldarera, Polyamine regulation of cyclic nucleotide enzymes, Ital. J. Biochem, (in press).

5. C. Clô, B. Tantini, C. Pignatti, C. Guarnieri, and C.M. Caldarera, Regulation of cyclic nucleotide metabolism by polyamines in heart cell cultures, Adv. Polyamine Res.,4:667 (1983).

6. C. Clô, C. Pignatti, B. Tantini, and C.M. Caldarera, A possible involvement of endogenous polyamines in the regulation of the steady-state levels of cyclic nucleotides in heart cell cultures Ital. J. Biochem.,31:437 (1982).

7. C. Clô, C. Pignatti, B. Tantini, S. Marmiroli, and C.M. Caldarera, Cyclic nucleotide enzymes in heart cell cultures treated with polyamine inhibitors, in "Polyamines in Biology and Medicine", C.M. Caldarera and U. Bachrach, eds., Clueb, Bologna (in press).

8. J. Jänne, H. Pösö, and A. Raina, Polyamines in rapid growth and cancer, Biochim. Biophys. Acta, 473:241 (1978).

9. U. Bachrach,   "The Function of Naturally Occurring Polyamines" Academic Press, New York (1973).

10. C. Clô, M.N. Coccolini, B. Tantini, and C.M. Caldarera, Increased sensitivity of heart cell cultures to norepinephrine after exposure to polyamine synthesis inhibitors, Life Sci.,27:67 (1980).

11. M.H. Bradford, A rapid and sensitive method for the quantitation of microgram quantities of protein utilizing the principle of protein-dye binding, Anal. Biochem.,72:248 (1976).

12. N.E. Newton, K. Ohno, and M.M. Adbel-Monem, Determination of diamines and polyamines in tissues by high-pressure liquid chromatography, J. Chromatogr.,124:277 (1976).

13. C. Pignatti, B. Tantini, C. Rossoni, E. Turchetto, and C. Clô, Involvement of polyamines in phospholipid-induced changes of cyclic nucleotide contents and macromolecular synthesis in cultured heart cells,Adv. Polyamine Res.,4:331 (1983).

14. P.S. Schönhöfer, and H.D. Peters, Role of cyclic nucleotides in cultured cells, in "Cyclic 3'-5' Nucleotides: Mechanism of Action", H. Cramer and J. Schultz, eds., John Wiley & Sons,

London, pp. 107-131 (1977).

15. K.P. Minneman, and L.L. Iversen, Eukephalin and opiate narcotics increase cyclic GMP accumulation in slices of rat nostriatum, Nature,262:313 (1976).

16. P.S. Rudland, N. Seeley, and W. Seifert, Cyclic GMP and cyclic AMP levels in normal and transformed fribroblasts, Nature,251: 417 (1974).

17. T.G. O'Brien, D. Saladik, and L. Diamond, Regulation of polyamine biosynthesis in normal and transformed hamster cells in culture, Biochim. Biophys. Acta,632:270 (1980).

18. W.L. Ryan and M.L. Heidrik, Role of cyclic nucleotides in cancer, Adv. Cyclic Nucleotide Res.,4:81 (1974).

19. V. Macchia, M.F. Neldolesi, and M. Chiariello, Adenyl-Cyclase in a transplantable thyroid tumor: loss of ability to respond to TSH, Endocrinology,90:1483 (1972).

20. T.N. Lincoln, and G.L. Vaughan, The role of adenosine 3',5'--monophosphate in the Transformation of Cloudman mouse melanoma cells, J. Cell. Physiol.,86:543 (1975).

21. W.B. Anderson, and I. Pastan, Altered adenylate cyclase activity: its role in growth regulation and malignant transformation of fibroblasts, Adv. Cyclic Nucleotide Res.,5:681 (1975).

22. M.S. Amer, Cyclic nucleotides in disease: on the biochemical etiology of hypertension, Life Sci.,17:1021 (1975).

23. R.W. Holley, Control of growth of mammalian cells in cell culture, Nature,258:487 (1975).

24. H.M. Temin, Studies on carcinogenesis by Avian Sarcoma viruses. VI. Differential multiplication of uninfected and of converted cells in response to insulin, J. Cell. Physiol.,69:377 (1967).

25. R. Dulbecco, Topoinhibition and serum requirement of transformed and untransformed cells, Nature,227:802 (1970).

IONIC SIGNAL TRANSDUCTION BY GROWTH FACTORS[*]

W.H. Moolenaar, L. Joosen, and S.W. de Laat

Hubrecht Laboratory
International Embryological Institute
Uppsalalaan 8
3584 CT Utrecht, the Netherlands

INTRODUCTION

The molecular mechanisms underlying the regulation of mammalian cell proliferation and differentiation are largely unknown. Insight into these complex mechanisms may be obtained by studying the mode of action of polypeptide growth factors, such as epidermal growth factor (EGF), platelet-derived growth factor (PDGF) and nerve growth factor (NGF). Growth factors, like all polypeptide hormones, initiate their action by binding to specific, high-affinity receptors on the cell surface. This interaction triggers a cascade of biochemical and physiological events in the cell, culminating in replicative DNA synthesis and cell division (EGF, PDGF, NGF), or in the expression of a differentiated phenotype (NGF). Understanding the mode of action of growth factors requires the identification of intracellular signals which are essential for the ultimate biological responses. In the search for growth factor-induced mitogenic signals much attention has been focussed on the earliest detectable cellular changes following the addition of growth factors to quiescent cells. Immediate consequences (< 1 min) of growth factor-receptor binding include tyrosine-specific protein phosphorylations[1-4], lipid break-down[5-7] and enhanced ion transport across the plasma membrane[8-14]. However, whether any of these 'immediate events' is causally related to the eventual initiation of DNA synthesis is still an open question.

The intention of the present chapter is to briefly review the ionic aspects of growth factor action. Emphasis will be put on the

[*] Research supported by the Netherlands Cancer Foundation (Koningin Wilhelmina Fonds).

role of electroneutral Na^+/H^+ exchange and cytoplasmic pH (pH_i) in
the mitogenic activation of quiescent mammalian cells. In addition,
we will summarize some of our recent findings on cytoplasmic free
Ca^{2+} as a potential primary messenger in the action of growth stimulants.

MITOGEN-INDUCED ELECTRICAL EVENTS

Most information on mitogen-induced ionic fluxes has been
gained from unidirectional tracer-flux studies. It seems appropriate
to recall that the magnitude of an electrodiffusional ion flux is
determined by (i) membrane permeability for that ion; (ii) transmembrane concentration gradient; and (iii) electrical membrane
potential. Thus, interpretations of hormone-induced ion fluxes are
of limited value unless the dynamic behaviour of the membrane
potential is known. Unfortunately, this basic fact is often overlooked in the biochemical and biological literature.

The electrical membrane potential (V_m), as measured with intracellular microelectrodes, is in turn determined by (i) the asymmetric
distribution of Na^+ and K^+ across the plasma membrane, and (ii) the
ratio of Na^+ to K^+ permeability. Rapid changes in ionic permeabilities
as induced by any surface stimulant manifest themselves as alterations
in V_m and membrane resistance. However, changes in the rate of
virtually electroneutral ion transporters (e.g. Na^+/H^+ exchange,
Na^+/K^+-ATPase) do not have an electrophysiological correlate, and
can only be revealed by an appropriate tracer-flux experiment. In
conclusion, microelectrode and tracer-flux studies should be done
in parallel to obtain a complete picture of hormone-dependent ion
movements.

The first direct demonstration that growth factors present in
serum elicit a dramatic change in ionic membrane properties of
quiescent cells was given by Hülser and Frank in 1971 (ref. 15).
These authors showed that growth-promoting serum protein induces an
immediate membrane depolarization (from $\sim$ -50 to $\sim$ -17 mV) in
embryonic rat fibroblasts, presumably due to an increase in monovalent ion permeability[15,16].

In 1979, we described a similar electrical phenomenon in serum-
-stimulated mouse neuroblastoma cells[17]. This rather complex serum
response could be resolved into two components each evoked by different serum components[11]. The growth-promoting serum fraction
induced a transient depolarizing phase, reminiscent of the serum
response observed by Hülser and Frank (Fig. 1A). In a subsequent
study[12], we have extended these initial observations to show that
serum rapidly depolarizes quiescent human fibroblasts (from -66 to
-15 mV), due to a rather unselective conductance increase for Na^+,K^+
and perhaps Ca^{2+} ions (Fig. 1B). A long-term depolarization is also

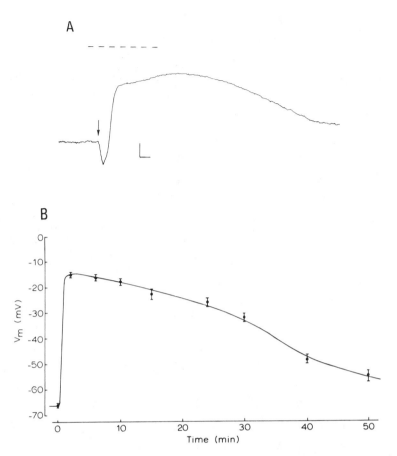

Fig. 1.   Transient membrane depolarizations during serum stimulation
of mouse neuroblastoma cells (A) and human fibroblasts (B)
as measured by intracellular microelectrodes. Dialyzed
serum (10-20%) was added at time zero (arrow in A). Cali-
bration bars in A: 10 sec (horizontal), 5 mV (vertical);
dashed line is zero voltage level. For full details see
refs. 11 and 12 respectively.

observed when quiescent fibroblasts are stimulated to grow after
'wound-formation' in a monolayer[18]. Although an intriguing phenomena
per se, similar changes in V_m are not evoked by mitogenic concentrations of EGF, insulin and NGF in various mammalian cells such as
fibroblasts[12], neuroblastoma[19,20] and pheochromocytoma[14] cells. In
an epithelial cell line, however, EGF has been reported to elicit a
small transient depolarization of unknown ionic origin[21].

Taken together, it appears that mitogen-induced changes in monovalent ion conductance and V_m are not a general phenomenon of growth-stimulation and it is thus legitimate to believe that V_m changes in itself are not essential for the generation of an intracellular mitogenic signal.

Na^+ AND K^+ FLUXES

A common event following mitogenic activation of quiescent cells is a rapid stimulation of the rate of the membrane-bound Na^+, K^+-ATPase (Na^+, K^+-'pump'), as evidenced by an increase in ouabain-sensitive K^+ or Rb^+ uptake[8,11,19,20,22-25]. The initial rate of the Na^+, K^+-pump increases at least two-fold within minutes of adding serum or polypeptide growth factors to quiescent cells. Also, the maturation factor NGF rapidly stimulates Na^+, K^+-pump activity in pheochromocytoma[14] and neuroblastoma[19] cells. In many cases, there is a ∿ 10-25% increase in the content of intracellular K^+, [8,11,22,23, 25], but this increase does not seem to be essential for entry into S phase in growth-stimulated fibroblasts[25]. At present, the prevailing opinion among most researchers is that changes in intracellular K^+ (and Na^+) concentrations do not serve, in themselves, as mitogenic signals in the action of growth stimulants[11,25-28].

Which mechanism underlies the stimulation of the Na^+, K^+-pump by growth factors? It is now realized that acceleration of the Na^+, K^+-pump is a direct consequence of a prior increase in Na^+ influx[9,11, 12,14,19,20,23] a large part of which is mediated by amiloride-sensitive Na^+/H^+ exchange. Smith and Rozengurt were the first to demonstrate that serum stimulation of mouse embryo fibroblasts leads to an increase in the rate of an amiloride-sensitive component of Li^+ uptake[29], but these authors did not interpret this result in terms of Na^+/H^+ exchange.

Strong evidence for the involvement of electroneutral Na^+/H^+ exchange in growth factor action came from our electrophysiological and tracer-flux experiments on mitogen-stimulated mouse neuroblastoma cells[8,19,20], human fibroblasts[12] and rat PC12 cells[14]. Comparison of the electrical data with Na^+ influx measurements led to the conclusion that mitogen-induced Na^+ influx was mediated, at least in part, by an electrically silent Na^+ pathway, which is distinct from the basal Na^+ permeability. This electroneutral Na^+ pathway was identified as a Na^+/H^+ exchange because of the following findings: (i) it is sensitive to amiloride, a known inhibitor of Na^+/H^+ exchange[30-32], while this drug does not affect the interaction between EGF and its receptor[12], and (ii) amiloride-sensitive Na^+ influx (and Na^+, K^+-pump activity) can be stimulated by simply exposing the cells to weak acids at constant external pH, suggesting that the Na^+ transporter is sensitive to cytoplasmic acid loads. Together, these results led to the proposal that growth factors

Table 1. Activation of Amiloride-sensitive Na$^+$ Influx
            by Various Mitogens

Mitogen	Cell Type	Reference
serum	mouse embryo fibroblasts	29
	mouse neuroblastoma cells	11
	human fibroblasts	12, 13
EGF	human fibroblasts	12, 13
	mouse neuroblastoma cells	19, 20
	rat PC12 cells	14
NGF	mouse neuroblastoma cells	19
	rat PC12 cells	14
PDGF	human fibroblasts	(unpublished data)
EGF + Glucagon + Insulin	rat hepatocytes	10
Thrombin	hamster fibroblasts	33
Lys-bradykinin	human fibroblasts	34

rapidly activate a previously 'quiescent' Na$^+$/H$^+$ exchanger. Increased Na$^+$ influx leads then to the observed stimulation of the Na$^+$,K$^+$ pump while it was predicted that the simultaneous efflux of H$^+$ would result in a rise in pH$_i$. The data in Table 1 indicate that activation of Na$^+$/H$^+$ exchange appears to be a common property of various growth stimulants.

The functioning of a Na$^+$/H$^+$ exchanger in the plasma membrane of mammalian cells and its role in pH$_i$ homeostasis is most readily assessed by continuously monitoring the kinetics of pH$_i$ recovery after an acute acid load[35]. Recent advances in pH$_i$-monitoring techniques, particularly the use of intracellularly trapped fluorescent indicators[36,37], have allowed us to obtain a fairly good understanding of the role of Na$^+$/H$^+$ exchange in pH$_i$ regulation in general and in the action of growth factors in particular[38,39]. First we will briefly summarize recent evidence on the role of Na$^+$/H$^+$ exchange in pH$_i$ homeostasis in normal non-excitable cells such as diploid human fibroblasts (HF cells).

Na^+/H^+ EXCHANGE IN pH_i REGULATION

Following an abrupt fall in pH_i, as induced by a NH_4^+-prepulse [35,40], pH_i rapidly recovers to its normal resting value of $\sim$ 7.05. This pH_i recovery follows an exponential time course and is mediated entirely by Na^+/H^+ exchange as indicated by the following findings: (i) pH_i recovery is accompanied by Na^+ uptake and net H^+ exit; (ii) pH_i recovery and concomitant Na^+,H^+ fluxes are reversibly blocked by amiloride; (iii) the rate of pH_i recovery depends critically on the external Na^+ concentration (half-maximal rate at $\sim$ 30 mM Na^+), but not on external anions like Cl^- and HCO_3^-; and (iv) Li^+ can substitute for Na^+ in pH_i recovery but other cations cannot. Figure 2 illustrates some of these aspects of pH_i regulation in HF cells (further details in ref. 38).

The exponentiality of pH_i recovery from an acid load is an important finding. As noted by others, working on pH_i regulation in large excitable cells[35,41], the net H^+ extrusion rate (J_{H^+}) may be written as

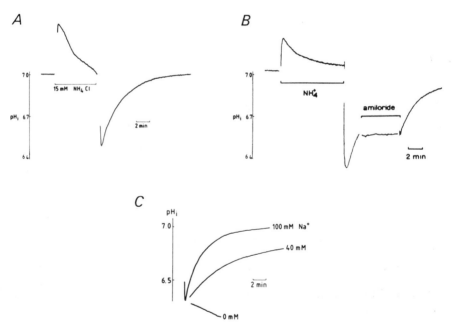

Fig. 2. *Fluorimetric recordings of the time course of pH_i recovery from an acid load induced by an NH_4Cl-prepulse in HF cells. Monolayers were loaded with bis(carboxyethyl)-carboxyfluorescein (BCECF) by uptake of its membrane-permeable ester. A, control. B, inhibition by 1 mM amiloride. C, dependence on external Na^+. Full details in ref. 38.*

$$J_{H^+} = \beta \ d(pH_i)/dt$$

where β is the cytoplasmic 'buffering power', defined as $d(OH^-$ added$)/$ $d(pH_i)$ and taken to be independent of pH_i[35,41]. Thus, in the absence of additional H^+ extruding mechanisms, the net rate of the Na^+/H^+ exchanger is proportional to $d(pH_i)/dt$. Exponential pH_i recovery then implies that the Na^+/H^+ exchange rate (J_{Na^+/H^+}) is linearly dependent on pH_i according to

$$J_{Na^+/H^+} = k\beta \ (pH_i' - pH_i)$$

where k is the rate constant of the exponential pH_i time course and pH_i' is a critical pH_i value ('threshold') at which Na^+/H^+ exchange activity apparently 'shuts off'[35,41]. It should be noted that the threshold pH_i does not correspond to thermodynamic equilibrium of the exchanger. Under normal physiological conditions there is still sufficient energy in the Na^+, H^+ gradients to raise pH_i nearly one pH unit more alkaline[35,39,41].

Aronson et al. have presented evidence that the pH_i-sensitivity of the Na^+/H^+ exchanger in renal microvilli is due to allosteric activation by internal H^+ at a site which is distinct from the internal transport site[42]. In contrast, external H^+ seems to compete with Na^+, Li^+ and amiloride for binding at an identical site according to the following sequence of binding affinities:

$$H^+ \gg \text{amiloride} \gg Na^+ \approx Li^+ \text{ (ref. 43).}$$

Such a model of competitive interaction would explain the observed sensitivity of pH_i to small changes in external pH in HF cells

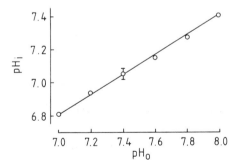

Fig. 3. Dependence of resting pH_i on external pH (pH_O) in quiescent HF cells. pH_O was shifted from 7.40 to the indicated values and pH_i was allowed to stabilize for at least 20 min. Fluorimetric measurements of pH_i as in Fig. 2.

(Fig. 3), but whether this model indeed holds for the Na^+/H^+ exchanger in intact fibroblasts remains to be clarified.

Na^+/H^+ exchange seems to be an ancient and ubiquitous ion transporter. It appears to be present not only in a wide variety of eukaryotic cells but also in bacteria[44] and mitochondria[45]. In animal cells Na^+/H^+ exchange has a primary role in pH_i regulation by virtue of its sensitivity to internal H^+. Its additional role in transmembrane signal transduction by growth factors will be discussed in the next section.

Na^+/H^+ EXCHANGE AND pH_i IN GROWTH FACTOR ACTION

If growth factors activate Na^+/H^+ exchange by some other mechanism than via intracellular acidification, then a rise in pH_i might be detectable. Using the conventional weak-acid distribution method, we and others[46,47] found that mitogenic stimulation of quiescent fibroblasts induces an amiloride-sensitive rise in intracellular pH of ~ 0.15 pH unit. A typical time course of the serum--induced pH_i shift in HF cells is shown in Figure 4. However, the weak-acid distribution method suffers from several technical limitations such as: (i) intracellular compartimentalization of labelled weak-acids; (ii) rapid loss of label during washing procedures; and (iii) sensitivity to changes in cellular volume. Consequently, interpretations based on this technique are of

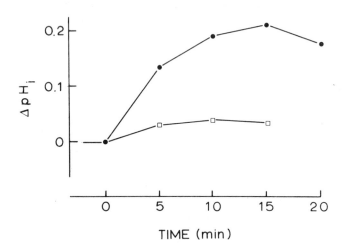

Fig. 4.   *Time course of change in pH_i in HF cells following addition of 10% serum (time zero) as inferred from changes in ^{14}C-DMO distribution in the absence (filled circles) and presence (open squares) of 0.5 mM amiloride.*

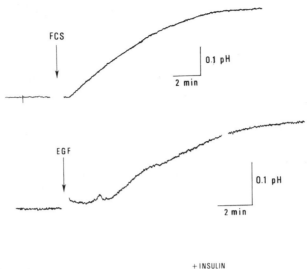

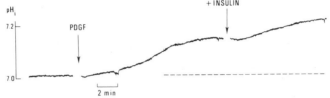

*Fig. 5.   Time course of pH$_i$ changes induced by 10% serum (FCS),
25 ng/ml EGF and 40 ng/ml PDGF respectively. Quiescent
cells were loaded with BCECF (ref. 38). Potentiating
effect of insulin (5 µg/ml) is shown in PDGF recording.
Human PDGF was a gift from C.-H. Heldin.*

limited value. We, therefore, have applied the above-mentioned
fluorimetric technique, using BCECF as a pH$_i$-sensitive indicator,
to obtain the first continuous pH$_i$ recordings of the cellular
response to growth stimulation[38].

Figure 5 shows typical examples of alkaline pH$_i$ shifts in HF
cells stimulated by serum (FCS), EGF and PDGF respectively.
Interestingly, insulin (5 µg/ml) markedly potentiates the effects
of EGF and PDGF, without affecting pH$_i$ when present alone[38]. In
general, the pH$_i$ shift is detectable within 10-20 seconds and is
virtually complete after 10-15 min. Table 2 summarizes the effects
of various growth stimulants on pH$_i$.

As expected, the mitogen-induced rise in pH$_i$ is inhibited by
amiloride or by Na$^+$ removal[38] and is accompanied by amiloride-
-sensitive Na$^+$ uptake[12,13]. Furthermore, the pH$_i$ response can be

Table 2.   Effects of Various Stimulants on pH_i

Stimulus	ΔpH_i (relative to control)
FCS (10%)	0.21 ± 0.01 (n = 9)
EGF (25 ng/ml)	0.09 ± 0.02 (4)
Insulin (5 µg/ml)	0.00 (12)
EGF + Insulin	0.15 ± 0.01 (6)
PDGF (40 ng/ml)	0.14 (2)
PDGF + Insulin	0.19 (2)

Quiescent HF cells were loaded with BCECF and assayed
for shifts in pH_i as detailed in Figs. 2 and 5 and in
ref. 38. The number of experiments is given in parentheses.

mimicked, at least qualitatively, by the Na^+/H^+ ionophore monensin
[39]. These results indicate that growth stimulants like EGF, PDGF
and serum rapidly activate the previously 'quiescent' Na^+/H^+
exchanger and thereby raise pH_i. Whether NGF similarly raises pH_i
in its target cells remains to be investigated. In the continuous
presence of growth factors, the elevated pH_i persists for at least
1-2 hours, whereas pH_i only slowly decays after mitogen washout;
actually, pH_i recovery is not complete until several hours after
EGF or FCS withdrawal. It thus appears that, while activation of
Na^+/H^+ exchange occurs rapidly, 'de-activation' of the exchanger is
a relatively slow process. Figure 6 summarizes the sequence of ionic
events following receptor occupancy.

Two major questions that now arise are: (i) by which mechanism(s)
do growth factors activate Na^+/H^+ exchange, and (ii) what is the
physiological significance of an early rise in pH_i? A clue as to
the first question was obtained by kinetic analysis of pH_i recovery
in both growth-stimulated and quiescent cells[38]. It appeared that
the overall rate of pH_i recovery from an NH_4^+-acid load remains
unaltered ($\sim$ 90% complete in 6-7 min) but that the value at which
pH_i stabilizes is $\sim$ 0.2 pH unit more alkaline. This is expressed
more quantitatively in Figure 7, where Na^+/H^+ exchange activity is
measured as $d(pH_i)/dt$. It is seen that growth stimulation simply
induces a shift of the function $d(pH_i)/dt$ versus pH_i to the right.
In other words, growth factors enhance the internal H^+ sensitivity
of the Na^+/H^+ exchanger. We have proposed[38] that this intrinsic
modification of the exchanger concerns the allosteric H^+ site at
the cytoplasmic face, resulting in an increased apparent affinity
for cytoplasmic H^+. It is tempting to speculate that such a conformational change is brought about by a mitogen-induced phosphorylation

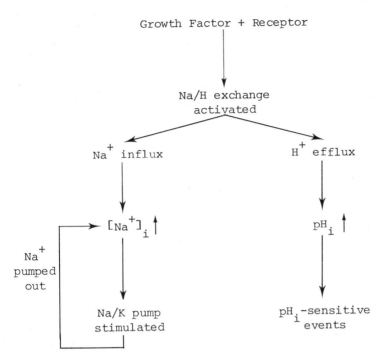

Fig. 6. *Scheme of ionic events following growth factor-receptor interaction. pH$_i$-sensitive events are discussed in the text.*

reaction. In fact, there is precedent for a possible link between phosphorylation of membrane proteins and the activation of carrier-mediated ion transport[48,49].

Several authors have claimed a regulatory role for cytoplasmic Ca^{2+} in stimulating Na^+/H^+ exchange[13,50,51]. This interpretation is mainly based on the finding that the Ca^{2+} ionophore A23187 may stimulate amiloride-sensitive Na^+ uptake in some cell types, including HF cells. However, A23187 (5 µg/ml) elicits an abrupt and persistent fall in pH$_i$[38]. Whatever the underlying mechanism of this cytoplasmic acidification, this result shows that there is no need to postulate a direct role for internal Ca^{2+} in the control of Na^+/H^+ exchange activity, since the apparent activation of Na^+/H^+ exchange can simply be explained by a prior fall in pH$_i$. It cannot, of course, be excluded that Ca^{2+} is involved in a more indirect way in the activation of Na^+/H^+ exchange, for example via calmodulin or a lipid breakdown product. Growth-stimulation studies using putative calmodulin antagonists and other lipid-interacting drugs lend support to this possibility[52-54].

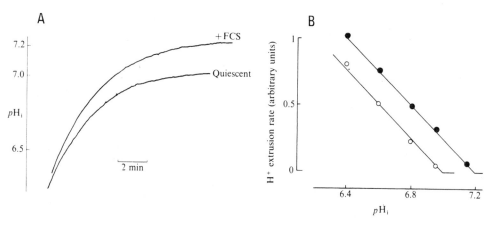

Fig. 7.   *A, superimposed recordings of exponential pH$_i$ recovery from*
a NH$_4^+$ acid load in quiescent and growth-stimulated HF
cells, respectively. After the first pH$_i$ recovery, 10% FCS
was added and 10 min later a NH$_4^+$ pulse was given. Recovery
of pH$_i$ from this second acid load was monitored in the
continuous presence of FCS. B, Relationship between net
rate of H$^+$ extrusion (that is Na$^+$/H$^+$ exchange activity)
and pH$_i$ of the experiment in A. Rate of H$^+$ extrusion
(arbitrary units) was measured as d(pH$_i$)/dt, estimated as
the net rise in pH$_i$ over 10-s intervals. (From ref. 38.)

Possible Physiological Effects of a Rise in pH$_i$

An early rise in pH$_i$ may be a fairly common response of metabolically 'dormant' cells to appropriate surface stimuli. It has
been described for fertilization of sea urchin and frog eggs[30,55],
the action of insulin on frog skeletal muscle[56], platelet activation
by thrombin[57] and for the chemotactic response of human neutrophils[58].
What may be the physiological relevance of this pH$_i$ shift? From
studies on various cell types it is known that a rise in pH$_i$ may
take part in the control of such diverse processes as protein
synthesis[59], cytoskeletal reorganization[60,61], cell-to-cell communication[62], oxygen consumption and cell motility[63,64]. Moore and
colleagues have presented strong evidence that the stimulation of
glycolosis by insulin in frog muscle is due to a rise in pH$_i$ mediated
by increased Na$^+$/H$^+$ exchange activity[65,66]. Indeed, the key rate-
-limiting enzyme of the glycolytic pathway, phosphofructokinase
(PFK), is extremely pH-sensitive[67]. It seems a plausible assumption
that physiological regulation of kinase activity by pH$_i$ is not
limited to the case of PFK. For example, Pouysségur et al. have
hypothesized that the amiloride-sensitive phosphorylation of ribosomal

Table 3.   Glycolytic Rate in HF Cells under Various
              Conditions Known to Alter pH_i

Conditions	Lactate production per 30 min (relative to control)
Control (pH 7.4)	1.0
Serum (10%)	1.58
EGF (20 ng/ml) + Insulin (5 µg/ml)	1.30
Monensin (2 µM)	1.83
pH 7.8	1.20
Amiloride (0.5 mM)	0.80
+ Serum	1.15
Na^+-free medium	0.55
+ Serum (Na^+-free)	1.30
+ Insulin + EGF	0.88

Aerobic lactate production was determined fluorimetrically using the Sigma lactate assay kit. The basal rate (control) corresponds to $\sim$ 15 µmole lactate/30 min/ 10^6 cells.

protein S6 is mediated by a rise in pH_i[33]. This claim has been questioned, however, on account of non-specific effects of amiloride on the activity of protein kinases[68].

We have tested the possibility that growth factor-dependent glycolosis may similarly be the consequence of an elevated pH_i. As shown in Table 3, artificially raising pH_i by either monensin or alkaline external pH stimulates aerobic lactate production in quiescent HF cells. However, addition of amiloride or removal of Na^+ does not affect the activation of glycolysis by serum or growth factors. One is therefore left with the conclusion that, although pH_i may modulate the glycolytic rate, mitogens utilize additional signals to activate glycolosis.

Ca^{2+} MOBILIZATION

Cytoplasmic free Ca^{2+}, $[Ca^{2+}]_i$, has been implicated as a 'second messenger' in the action of various hormones, but little is known about the dynamic behaviour of $[Ca^{2+}]_i$ in quiescent cells stimulated to reinitiate DNA synthesis and cell division. It has been shown that serum stimulation of quiescent fibroblasts induces rapid changes in $^{45}Ca^{2+}$ fluxes[69,70]. Furthermore, indirect evidence suggests that EGF may cause a redistribution of cytoplasmic Ca^{2+} in its target cell[5,71] but changes in $[Ca^{2+}]_i$ have not previously been detectable.

Using the novel fluorescent Ca^{2+} indicator 'Quin 2', Tsien and co-workers have recently described an early 2-3 fold rise in $[Ca^{2+}]_i$ in lectin-stimulated lymphocytes[72]. We have used the same technique to show for the first time that serum stimulation of quiescent HF cells rapidly induces a $\sim$ 2-fold rise in $[Ca^{2+}]_i$ (Fig. 8). The increase in $[Ca^{2+}]_i$ is detectable within 1-2 seconds, it peaks at $\sim$ 15 seconds, and then gradually decays over the next few minutes to stabilize at a level which is $\sim$ 20% above the normal resting value of $\sim$ 0.1 μM. The serum-induced Ca^{2+} transient is not significantly attenuated in Ca^{2+}-free media containing 0.5 mM EGTA, indicating that transmembrane Ca^{2+} influx does not participate in this response. We therefore conclude that Ca^{2+} is liberated from internal stores such as plasma membrane, mitochondria or endoplasmatic reticulum.

These results strongly suggests that Ca^{2+} may be a primary messenger in the action of growth stimulants. Mobilized Ca^{2+} interacts with calmodulin to activate various enzyme reactions and, in concert with lipid breakdown, may activate the novel protein kinase C described by Nishizuka and co-workers[73]. Whether any of these Ca^{2+}-dependent pathways is involved in the activation of Na^+/H^+ exchange is an important area for future research.

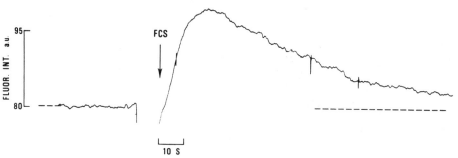

Fig. 8.   *Fluorimetric recording of Ca^{2+}-dependent Quin 2 fluorescence following serum stimulation of HF cells. HF monolayers were loaded with Quin 2 by uptake of its membrane-permeable ester (50 μM; 30 min incubation at 37°C). Basal fluorescence (arbitrary units) corresponds to about 0.1 μM free Ca^{2+}. Peak transient represents an $\sim$ 2-fold increase in $[Ca^{2+}]_i$. Calibration was done by the digitonin/EGTA method (ref. 72).*

CONCLUDING REMARKS

Various molecular events have been identified over the past few years which may participate in transmembrane signalling by growth factors. These include tyrosine-specific protein phosphorylation, lipid breakdown and two major ionic events discussed above, namely elevation of cytosolic free Ca^{2+} and activation of Na^+/H^+ exchange with a resultant rise in pH_i. Uncertainty exists about the inter-relationship, if any, between these immediate events and about the relevance of each of these steps in the eventual initiation of DNA synthesis. Selective pharmacological inhibitors, without non-specific side-effects, would be of great help in elucidating these questions, but such agents are lacking at present. It is hoped that the use of various monoclonal antibodies against growth factor receptors, Na^+/H^+ exchanger etc. will greatly facilitate this problem. Another challenge for future investigations is to test whether the molecular events induced by growth factors have their correlates in the action of transforming oncogene products, some of which are known to be tyrosine-specific protein kinases. Such an approach would offer a unique opportunity to study the overlap between 'normal' growth factor action and malignant growth induced by altered oncogene expression.

REFERENCES

1. G. Carpenter, L. King, and S. Cohen, J. Biol. Chem. 254: 4884 (1979).
2. T. Hunter and J.A. Cooper, Cell 24: 741 (1981).
3. J.A. Fernandez-Pol, J. Biol. Chem. 256: 9742 (1981).
4. B. Ek and C.-H. Heldin, J. Biol. Chem. 257: 10486 (1982).
5. S.T. Sawyer and S. Cohen, Biochemistry 20: 6280 (1981).
6. A. Habenicht, J. Glomset, W.C. King, C. Nist, C.D. Mitchell, and R. Ross, J. Biol. Chem. 256: 12329 (1981).
7. W.T. Shier and J.P. Durkin, J. Cell. Physiol. 112: 171 (1982).
8. E. Rozengurt and L.A. Heppel, Proc. Natl. Acad. Sci. USA 72: 4492 (1975).
9. J.B. Smith and E. Rozengurt, Proc. Natl. Acad. Sci. USA 75: 5560 (1978).
10. K.S. Koch and H.L. Leffert, Cell 18: 153 (1979).
11. W.H. Moolenaar, C.L. Mummery, P.T. van der Saag, and S.W. de Laat, Cell 23: 789 (1981).
12. W.H. Moolenaar, Y. Yarden, S.W. de Laat, and J. Schlessinger, J. Biol. Chem. 257: 8502 (1982).
13. M.L. Villereal, J. Cell. Physiol. 107: 359 (1981).
14. J. Boonstra, W.H. Moolenaar, Ph.H. Harrison, P. Moed, P.T. van der Saag, and S.W. de Laat, J. Cell Biol. 97: 92 (1983).
15. D.F. Hülser and W. Frank, Z. Naturforsch. 26B: 1045 (1971).
16. W. Frank, H. Ristow, R. Hoffman, J. Grim, and J. Veser, Adv. Metab. Disord. 8: 337 (1975).

17. W.H. Moolenaar, S.W. de Laat, and P.T. van der Saag, Nature 279: 721 (1979).
18. C.D. Cone and M. Tongier, J. Cell. Physiol. 82: 373 (1973).
19. W.H. Moolenaar, C.L. Mummery, P.T. van der Saag, and S.W. de Laat, in: "Membranes in Tumour Growth", T. Galeotti et al., eds., Elsevier, Amsterdam (1982).
20. C.L. Mummery, P.T. van der Saag, and S.W. de Laat, J. Cell. Biochem. 21: 63 (1983).
21. P. Rothenberg, L. Reuss, and L. Glaser, Proc. Natl. Acad. Sci. USA 79: 7783 (1982).
22. J.T. Tupper, F. Zorgniotti, and B. Mills, J. Cell. Physiol. 91: 429 (1977).
23. S.A. Mendoza, N.M. Wigglesworth, P. Pohjanpelto, and E. Rozengurt, J. Cell. Physiol. 103: 17 (1980).
24. J.G. Kaplan, Ann. Rev. Physiol. 40: 19 (1978).
25. C.N. Frantz, D.G. Nathan, and C.D. Scher, J. Cell Biol. 88: 51 (1981).
26. J.T. Tupper and L. Zografos, Biochem. J. 174: 1063 (1978).
27. D. Moscatelli, H. Sanui, and A.H. Rubin, J. Cell. Physiol. 101: 117 (1979).
28. B.J. Hazelton and J.T. Tupper, Biochem. J. 194: 707 (1981).
29. J.B. Smith and E. Rozengurt, J. Cell. Physiol. 97: 441 (1978).
30. J.D. Johnson, D. Epel, and M. Paul, Nature 262: 661 (1976).
31. J.L. Kinsella and P.S. Aronson, Am. J. Physiol. 241: F374 (1981).
32. W.H. Moolenaar, J. Boonstra, P.T. van der Saag, and S.W. de Laat, J. Biol. Chem. 256: 12883 (1981).
33. J. Pouysségur, J.C. Chambard, A. Franchi, S. Paris, and E. van Obberghen-Schilling, Proc. Natl. Acad. Sci. USA 79: 7778 (1982).
34. N.E. Owen and M.L. Villereal, Cell 32: 979 (1983).
35. A. Roos and W.F. Boron, Physiol. Rev. 61: 296 (1981).
36. J.A. Thomas, R.N. Buchsbaum, A. Zimniak, and E. Racker, Biochemistry 18: 2210 (1979).
37. T.J. Rink, R.Y. Tsien, and T. Pozzan, J. Cell Biol. 95: 189 (1982).
38. W.H. Moolenaar, R.Y. Tsien, P.T. van der Saag, and S.W. de Laat, Nature 304: 645 (1983).
39. W.H. Moolenaar, L. Joosen, L.G.J. Tertoolen, and S.W. de Laat, J. Biol. Chem. (submitted).
40. W.F. Boron and P. de Weer, J. Gen. Physiol. 67: 91 (1976).
41. W.F. Boron, J. Membr. Biol. 72: 1 (1983).
42. P.S. Aronson, J. Nee, and M. Suhm, Nature 299: 161 (1982).
43. P.S. Aronson, M. Suhm, and J. Nee, J. Biol. Chem. 258: 6767 (1983).
44. I.C. West and P. Mitchell, Biochem. J. 144: 87 (1974).
45. P. Mitchell and J. Moyle, Eur. J. Biochem. 9: 149 (1969).
46. W.H. Moolenaar, S.W. de Laat, C.L. Mummery, and P.T. van der Saag, in: "Ions, Cell Proliferation and Cancer", 151, A.L. Boynton et al., eds., Academic Press, New York (1982).
47. S. Schuldiner and E. Rozengurt, Proc. Natl. Acad. Sci. USA 79: 7778 (1982).

48. S.A. Rudolph and P. Greengard, J. Biol. Chem. 249: 5684 (1974).
49. H.C. Palfrey, S.L. Alper, and P. Greengard, J. Exp. Biol. 89: 103 (1980).
50. M. Taub and M.H. Saier, J. Biol. Chem. 254: 11440 (1979).
51. D.J. Benos, Am. J. Physiol. 242: C131 (1982).
52. N.E. Owen and M.L. Villereal, Proc. Natl. Acad. Sci. USA 79: 3537 (1982).
53. N.E. Owen and M.L. Villereal, Biochem. Biophys. Res. Commun. 109: 762 (1982).
54. N.E. Owen and M.L. Villereal, Exp. Cell Res. 143: 37 (1983).
55. R. Nuccitelli, D.J. Webb, S.T. Lagier, and G.B. Matson, Proc. Natl. Acad. Sci. USA 78: 4421 (1981).
56. R.D. Moore, Biochem. Biophys. Res. Commun. 91: 900 (1979).
57. W.C. Horne, N.E. Norman, D.B. Schwartz, and E.R. Simons, Eur. J. Biochem. 120: 295 (1981).
58. T. Molski, P. Naccache, M. Volpi, L. Wolpert, and R. Sha'afi, Biochem. Biophys. Res. Commun. 94: 508 (1980).
59. J.L. Grainger, M.M. Winkler, S.S. Shen, and R.A. Steinhardt, Devl. Biol. 68: 396 (1979).
60. D.A. Begg and L.A. Rebhun, J. Cell Biol. 83: 241 (1979).
61. C.S. Regula, J.R. Pfeiffer, and R.D. Berlin, J. Cell Biol. 89: 45 (1981).
62. D.C. Spray, A.L. Harris, and M.V.L. Bennet, Science 211: 712 (1981).
63. R. Christen, R.W. Schackmann, and B.M. Shapiro, J. Biol. Chem. 257: 14881 (1982).
64. H.C. Lee, C. Johnson, and D. Epel, Devl. Biol. 95: 31 (1983).
65. R.D. Moore, M.L. Fidelman, and S.H. Seeholzer, Biochem. Biophys. Res. Commun. 91: 905 (1979).
66. M.L. Fidelman, S.H. Seeholzer, K.B. Walsh, and R.D. Moore, Am. J. Physiol. 242: C87 (1982).
67. B. Trivedi and W.H. Danforth, J. Biol. Chem. 241: 4110 (1966).
68. R. Holland, J.R. Woodgett, and D.G. Hardie, FEBS Letters 154: 269 (1983).
69. J.T. Tupper, M. Del Rosso, B. Hazelton, and F. Zorgniotti, J. Cell. Physiol. 95: 71 (1978).
70. A. Lopez-Rivas and E. Rozengurt, Biochem. Biophys. Res. Commun. 114: 240 (1983).
71. P. Sherline and R. Mascardo, J. Cell Biol. 95: 316 (1982).
72. R.Y. Tsien, T. Pozzan, and T.J. Rink, Nature 295: 68 (1982).
73. Y. Nishizuka, Trends Biochem. Sci. 8: 13 (1983).

EFFECT OF A CONDITIONED MEDIUM ON THE GROWTH RATE

AND NUTRIENT TRANSPORT IN SV3T3 CELLS

Lucia Silvotti, Pier Giorgio Petronini,
Giuseppe Piedimonte and Angelo F. Borghetti

Istituto di Patologia Generale, Università di Parma
Via Gramsci 14, 43100 Parma, Italy

Growth controls of transformed cells are in general relaxed but not absent; that is,they can grow under some culture conditions that are known to block the growth of non transformed cells[1]. For instance, among these culture conditions cell density is a well known phenomenon that controls the growth of 'normal' 3T3 cells[2]. In simian virus 40-transformed 3T3 cells (SV3T3) we have recently shown that the proliferation rate is not strictly correlated with cell density[3]. Sparse cultures exhibited marked fluctuations in their growth rate whereas, under comparable conditions, crowded cultures retained some form of growth control by density. Furthermore, several factors other than cell density appeared to play a role in modulating the proliferation rate of sparse SV3T3 cells. These factors are likely to include the number of cells initially plated, the time spent by the cells in the same culture conditions, the production of autocrine growth factors and the concentration of the serum in the culture medium. It is known that serum concentration primarily affects the saturation density at which a population of non-transformed 3T3 cells becomes quiescent[4]. In contrast, the saturation density of SV3T3 cells is much less affected by serum; furthermore, these cells can grow significantly at low serum concentration[5].

The relationship between nutrient transport rate and cell growth is less clear. An increased membrane permeability for low molecular weight nutrients has been proposed as an essential event for optimal cell proliferation[6]. Moreover, it has been indicated that an alteration in membrane transport of critical metabolites may result in differential growth rates of malignant cells.[6,7] However, some

233

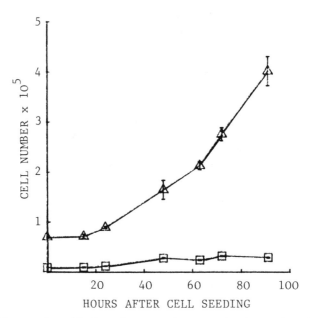

Fig. 1.   Effect of cell density on the population kinetics in SV3T3
cell culture grown at 0.5% serum.   The number of cells
initially plated were: $\triangle$, 0.8 x 10^4 cells/cm^2 and $\square$,
1.1 x 10^3 cells /cm^2.

reports have contradicted this suggestion[8-10].

Here we report preliminary results on the effect of cell density
on the proliferation rate of transformed SV3T3 cells when grown at
low serum concentration.   Moreover, the activity of a conditioned
medium on the growth rate and amino acid transport of SV3T3 cells
is also reported.

Details of cell culturing and amino acid transport assay have
been described in a previous paper[3].   Conditioned medium (CM) is
referred to a culture medium at low (0.5%) serum concentration in
which a high cell density population of SV3T3 (10^5 cells/cm^2) has
grown for a fixed period of time (40 hrs).   For cell counting,
SV3T3 cells were detached by trypsinization from the substratum,
and an aliquot of the resulting cell suspension was counted after
proper dilution in a Coulter counter, model ZM.   The values were
automatically corrected for coincidence counting and averaged.

As shown in figure 1, the density of the cells seeded affects
the proliferation rate when SV3T3 are grown at low serum concentration.   This result indicates that cells seeded at a density of
0.8 x 10^4 cells / cm^2 autostimulate their growth in the presence

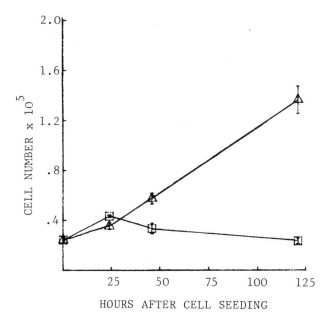

Fig. 2.   Effect of conditioned medium (CM) on the population
          kinetics in SV3T3 cell culture grown at 0.5% serum.
          □ , control; △ , in the presence of CM.

of only traces of serum, whereas low density cultures grow at a very
low rate. Assuming that this difference in the proliferation rate
may be due to a higher level of autocrine factors secreted by the
high density culture in comparison to the low density culture, we
tested the effect of a conditioned medium on the growth of SV3T3
cells seeded at low density and in the presence of 0.5% serum. As
shown in figure 2 CM appears to stimulate the proliferation rate
after a lag of about 30 hours in a highly significative way. The
addition of CM to sister cultures led to an increased uptake of
proline (see table 1). It should be noted that this transport
stimulation appeared to be concomitant with the changes of the cell
growth, as expressed by the increases of the protein content per
dish. From the results presented in the same table, one can see
that cell cultures seeded at a little higher inoculum poorly respond
to the presence of CM as regards to the growth rate or the amino
acid transport activity.

     Taken together, these results are in agreement with our previous
observations on the peculiar characteristics of growth of transformed
cells[3]. At low serum concentration SV3T3 cells are still capable
of growing even if at low rate. In this culture condition the cell

Table 1. Effect of conditioned medium on proline transport
          in SV3T3 cells

Number of cells seeded x cm^2	Cell density obtained	Proline uptake	% difference vs. control
2.3×10^3	4.47 ± 0.17	0.22 ± 0.02	-
2.3×10^3 + CM	9.97 ± 1.24	0.45 ± 0.02	+ 105
5.1×10^3	8.18 ± 1.57	0.22 ± 0.03	-
5.1×10^3 + CM	11.44 ± 2.20	0.25 ± 0.05	+ 14

Cells were seeded 48 hrs before transport measurements.
Initial rates of proline uptake (1 min) were determined at
0.1 mM substrate concentration. Cell density is expressed
as ug of protein/cm^2 and proline uptake as umoles/min/ml of
intracellular water. Values are the mean $\pm$ S.D. of three
independent determinations.

density modulates the proliferation rate. The replacement of only
one-third or less of the normal growth medium with CM was found to
be adequate to increase the growth rate in low density cultures,
whereas small or none effects occur in subconfluent or crowded
cultures (our unpublished observations). This growth behaviour of
low density cultures when cultivated at low serum concentration
and their response to the addition of CM, indicates that SV3T3 cells
secrete growth factors that stimulate their own growth.

One of the early events which occurs after exposure of the
cells to mitogenic stimuli is the increase in the rate of uptake of
low molecular weight substances: thus the rate of transport of
nutrients has been proposed as a possible mechanism by which cell
proliferation is affected[6]. Serum has been considered one of the
best stimuli for cell proliferation and the concomitant increase
of nutrient uptake[2], including amino acids[11]. However, it should
be noted that several factors or conditions, including cell density,
are effective in the control of nutrient transport[12,13]. In the
experiments above described, we took advantage of a model, the
transformed cell line SV3T3, which can grow almost independently of
the serum presence, and of a culture condition, the serum limitation,
that reduces or abolishes the density control of amino acid uptake
(our unpublished results). Within the limits used in this analysis
(low serum concentration and cell culturing at low density) CM

stimulates amino acid transport. This effect appears to be restricted to the amino acid transport system A and is correlated with the
increase in the growth rate. Whether in SV3T3 cells the amino acid
uptake increase is a normal part of the pleiotypic proliferative
programme rather that an irrelevant change coincidentally caused by
the interaction with the cell membrane of presumptive 'growth
factors' present in the conditioned medium is currently under
investigation.

ACKNOWLEDGEMENTS

    This investigation was supported by the Consiglio Nazionale
delle Ricerche and by Ministero della Pubblica Istruzione, Rome, Italy.

REFERENCES

1. A. B. Pardee, J. Campisi, and R. G. Croy, Differences in
    growth regulation of normal and tumor cells, in: "Cell
    proliferation, cancer and cancer therapy," R. Baserga,
    ed., The New York Academy of Science, New York (1982).
2. R. W. Holley, Serum factors and growth control, in: "Control
    of proliferation in animal cells", B. Clarkson and R.
    Baserga, eds., Cold Spring Harbour Laboratory, Cold
    Spring (1974).
3. G. Piedimonte, A. F. Borghetti, and G. G. Guidotti, Effect
    of cell density on growth rate and amino acid transport
    in simian virus 40-transformed 3T3 cells, Cancer Res.
    42:4690 (1982).
4. R. W. Holley and J. A. Kiernan, "Contact inhibition" of
    cell division in 3T3 cells, Proc.Nat.Acad.Sci.
    (U.S.A.) 60:300 (1968).
5. J. C. Bartholomew, H. Yokota, and P. Ross, Effect of serum
    on the growth of balb 3T3 A31 mouse fibroblasts and an
    SV40-transformed derivative, J.Cell.Physiol. 88:277
    (1976).
6. A. B. Pardee, Cell division and a hypothesis of cancer,
    Natl.Cancer Monogr. 14:7 (1964).
7. R. W. Holley, A unifying hypothesis concerning the nature
    of malignant growth, Proc.Nat.Acad.Sci.(U.S.A.) 69:2840
    (1972).
8. H. Bush and M. Shodell, Uptake of low molecular weight sub-
    stances by SV40-transformed 3T3 cells is invariant with
    growth rates in the presence and absence of serum,

Exp.Cell Res. 114:27 (1978).

9. J. H. Robinson and J. A. Smith, Density-dependent regula-
    tion of proliferation rate in cultured, androgen-re-
    sponsive , tumour cells, J.Cell.Physiol. 89:111 (1976).

10. C. R. Trash and D. D. Cunningham, Dissociation of increased
    hexose transport from initiation of fibroblast prolif-
    eration, Nature 252:45 (1974).

11. M. Tramacere, A. F. Borghetti, and G. G. Guidotti, Serum-
    -mediated regulation of amino acid transport in cultured
    chick embryo fibroblasts, J.Cell.Physiol. 93:425 (1977).

12. A. F. Borghetti, G. Piedimonte, M. Tramacere, A. Severini,
    P. Ghiringhelli, and G.G.Guidotti, Cell density and
    amino acid transport in 3T3, SV3T3, and SV3T3 revertant
    cells, J.Cell.Physiol. 105:39 (1980).

13. G. G. Guidotti, A. F. Borghetti, and G. C. Gazzola, The
    regulation of amino acid transport in animal cells,
    Biochim.Biophys.Acta 515:329 (1978).

OSMOREGULATION OF GROWTH RATE AND NUTRIENT TRANSPORT

IN CULTURED FIBROBLASTS

Pier Giorgio Petronini, Mariarosaria Tramacere and
Angelo F. Borghetti

Istituto di Patologia Generale, Università di Parma
Via Gramsci 14, 43100 Parma, Italy

Microorganisms, plants and eukaryotic cells are able to grow
under a variety of environmental conditions, including low and/or
high temperature, availability of water, concentration of salts,
light intensity etc.[1]. Recently we have investigated one of the
mechanisms of cellular adaptation to osmotic stress, id.è., the
alteration in activity of nutrient transport processes[2]. In
hyperosmolarity-treated cultured fibroblasts these changes consist
of an increase in transport activity for neutral amino acids and
appear to be restricted to the A system of mediation. They require
a continuous exposure of the cells to a hyperosmolar medium and
their occurence is prevented by the presence in the culture medium
of inhibitors of RNA and protein synthesis[2].

Here we report results of further investigation on the mechanism of this adaptive change of membrane transport and its correlation with the growth characteristics of the cell culture.

Details of cell culturing, hyperosmolar treatment and amino
acid transport assay have been reported in a previous paper[2]. For
cell counting, monolayers of chick embryo fibroblasts (CEFs) were
detached by trypsinization from the substratum, and an aliquot of
the resulting cell suspension was counted in a Coulter counter,
model ZM. The values were automatically corrected for coincidence
counting and averaged. The role of intracellular Na^+ level on the
activity of amino acid transport System A was explored. In
hyperosmolarity-treated cells the initial velocity of proline
transport as a function of cell depletion in osmolar, Na^+-containing
medium was determined (see figure 1). In these cells a depletion

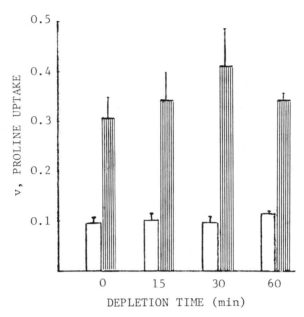

Fig. 1. Effect of the depletion time on the initial rate (1 min)
         of proline uptake, measured at 0.05 mM substrate concentra-
         tion, in normal (☐) or hyperosmolarity-treated (▥) cells.
         Values are expressed as umoles/min/ml of intracellular water
         and are the mean ± S.D. of three independent determinations.

Table 1. Effect of depletion time on intracellular Na$^+$
          in hyperosmolarity-treated cells

	Intracellular Na$^+$	
Depletion time	control	treated cells ·
--	25 + 1.2 mM	41 + 4.5 mM
15 min	25 + 1.4 mM	27 + 3.4 mM
30 min		24 + 3.4 mM
45 min	26 + 1.5 mM	27 + 2.8 mM

72 hrs after seeding CEF cultures were shifted to medium
of normal (143 mM) or altered (200 mM) Na$^+$. After additional 4 hrs cultures were processed to determine the
intracellular Na$^+$ by atomic absorption spectrometry
following a depletion phase in osmolar Earle's solution.

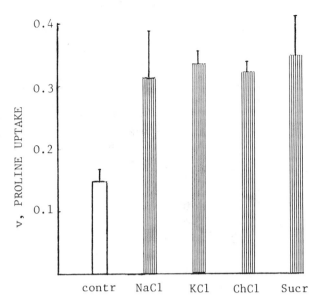

Fig. 2. Effect of the hyperosmolarity induced by Na^+, K^+, choline
         or sucrose on proline transport. For details see the
         legend of figure 1.

step which reduces the internal Na^+ to physiological levels (see
table 1) has not appreciably changed the increased amino acid
transport.

The possibility that the observed changes in amino acid transport were due to osmotic properties of the medium and not a specific
effect of Na^+ as a cation was then examined. Figure 2 shows that
an alteration of proline transport similar to that seen with the
high Na^+ treatment was obtained by adding in excess to the culture
medium such ions as K^+ or choline and carbohydrates such as sucrose.

In order to determine the effect of hyperosmolar treatment on
population kinetics of CEFs, the behaviour of the cell number was
determined as a function of time following the addition of the salt
excess. As presented in figure 3, hyperosmolarity-treated cells
proliferate at a slightly reduced rate in comparison to control
cells.

Taken together, these results confirm and extend our previous
observation on the osmoregulation of amino acid transport in CEFs.
Here we have presented evidence that this adaptive nutrient transport change is not dependent by the internal level of Na^+ and is
not related to the type of molecules involved in the alteration of

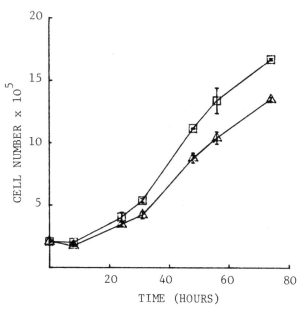

Fig. 3.   Growth rate of normal (▢) or hyperosmolarity-treated (△) cells. The alteration in osmolarity occurred 6 hrs after cell seeding.

the medium.   This result indicates that the cells are ready to monitor the osmotic strength of the environment rather than a Na^+ specific signal.

The results on the proliferation rate in hypertonicity-treated cells indicate that the increase in transport activity is not accompanied by parallel changes in the rate of growth. A lack of correlation between changes in nutrient transport and rate of cell growth has been previously reported by other workers[3,4] and by our group[5] in virus-transformed and in neoplastic cells. However, if cell proliferation is evaluated by the growth rate quotient, an estimate which improves the accuracy of cell growth measurements[5], none difference between treated and untreated cultures results (data not presented here). This outcome suggests that the lower steepness observed in the growth curve of hyperosmolarity-treated cells in comparison to control might be related to a reduced number of proliferating cells rather than to a different growth behaviour.

An additional question raised by the present investigation is the role of this regulatory aspect of amino acid transport in the physiology of a animal cell which will probably never face a significant change in osmolarity. However, the importance of cellular

adaptations to osmotic stress is well known in prokaryotes[6,7]. For
instance, non-halophilic bacteria have the availability to balance
the environmental osmotic pressure by regulating intracellular ionic
osmolarity and by intracellular accumulation of amino acids[8].
Indeed to prevent the harmful effects of ionic stress, bacteria are
known to accomplish this osmotic balance by derepressing inducible
transport systems for K^+ and by a build-up of the internal concentrations of amino acids such as proline and/or glutamate. Why an
animal cell should maintain the ability to respond to a change in
the enviromental osmolarity which it will never experience is a
matter of speculation. Indeed, in higher animals, homeostatic
mechanisms evolved to maintain a controlled osmolarity: only in
severe pathological conditions such as hyperosmolar coma is a significant derangement from balanced osmolarity known to occur. As
recently suggested by Spring and Ericson[9], the importance of having
membrane transport processes that are modulated may relate not only
to the control of cell osmolarity and/or cell volume but to the
transduction of a variety of signals such as those given by hormones,
growth factors, nutrient supplies etc. Thus, when animal cells are
exposed to hyperosmolarity an ancestral mechanism usually involved
in the transduction of a specific signal may be incidentally triggered.

Osmotic stress is an important environmental signal for bacteria,
plants and invertebrate animals. In this connection the well
known adaptive properties of amino acid transport system A in animal
cells, its activity being modulated by a number of factors and environmental conditions[10], add support to the above suggestion of a
newly acquired role of membrane transport systems in animal cells
in enabling them to respond when they are suddenly exposed to a
modified environment.

ACKNOWLEDGEMENTS

This investigation was supported by the Consiglio Nazionale
delle Ricerche and by Ministero della Pubblica Istruzione, Rome, Italy.

REFERENCES

1. P. J. Quinn, Models for adaptive changes in cell membranes,
    Biochem.Soc.Trans. 11:329 (1983).

2. M. Tramacere, P. G. Petronini, A. Severini, and A. F. Bor-
    ghetti, Osmoregulation of amino acid transport activity
    in cultured fibroblasts, Exp.Cell Res. in press (1983).

3. J. H. Robinson and J. A. Smith, Density-dependent regulation of proliferation rate in cultured, androgen-responsive, tumour cells, J.Cell.Physiol. 89:111 (1976).

4. H. Bush and M. Shodell, Uptake of low molecular weight substances by SV40-transformed 3T3 cells is invariant with growth rates in the presence and absence of serum, Exp. Cell Res. 114:27 (1978).

5. G. Piedimonte, A. F. Borghetti, and G. G. Guidotti, Effect of cell density on growth rate and amino acid transport in simian virus 40-transformed 3T3 cells, Cancer Res. 42:4690 (1982).

6. W. Epstein and L. Laimins, Potassium transport in Escherichia coli: diverse systems with common control by osmotic forces, Trends Biochem.Sci. 5:21 (1980).

7. D. Le Rudulier and R. C. Valentine, Genetic engineering in agriculture: osmoregulation, Trends Biochem.Sci. 7:431 (1982).

8. J. C. Measures, Role of amino acids in osmoregulation of non-halophilic bacteria, Nature 257:398 (1975).

9. K. R. Spring and A. C. Ericson, Epithelial cell volume modulation and regulation, J.Membrane Biol. 69:167 (1982).

10. M. Tramacere, A. F. Borghetti, and G. G. Guidotti, Serum-mediated regulation of amino acid transport in cultured chick embryo fibroblasts, J.Cell.Physiol. 93:425 (1977).

REGULATION OF THE UROKINASE mRNA SYNTHESIS BY TUMOR PROMOTERS

AND EPIDERMAL GROWTH FACTOR

M. Patrizia Stoppelli, P. Verde, Patrizia Galeffi,
Elsbjeta Kajtaniak Locatelli and F. Blasi

International Institute of Genetics and Biophysics
CNR, via Marconi 10, 80123 Naples, Italy

INTRODUCTION

Plasminogen activator is a serine protease that activates
its substrate plasminogen to plasmin. This process provides the
cell with a tool for regulating the availability of a very active
extracellular protease, plasmin, and thus to finely regulate extracellular proteolysis. Fibrinolysis is an example of such reactions.
Control of extracellular proteolysis, however, is required in various
physiological and pathological events, like cells and organ differentiation, tumor formation, response to inflammatory stimuli, DNA
damage, etc. (refs. 1 and 2 contain extensive citations of previous
work). Several agents have been reported to finely regulate plasminogen activator levels: among these, transformation with tumor viruses
(3-5), retinoic acid (6), tumor promoters (7,9), hormones (1) and
growth factors (10-12).

Two forms of plasminogen activators have been characterized
in the human species: the endothelial and the urinary forms, immuno-
logically unrelated. The endothelial form, tissue activator, has
a molecular weight of 70.000 (13). Its sequence has been deduced
by the nucleotide sequence of the cloned cDNA (14). The urinary
activator, urokinase, has a molecular weight of 54,000 daltons,
has been isolated from human urines and found to consist of two
chains, the A (18,000 daltons) and the B chain (30,000 daltons),
linked together by disulfide bonds. The amino acid sequence of the
two chains has been determined (15-17). The structure of urokinase
is depicted in Fig. 1 in a simplified form.

245

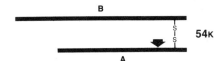

Fig. 1 Schematic representation of the structure of urinary high
m.w. urokinase. A (18,000 daltons) and the B chain (30,000
daltons) are held together by disulfide bonds. The enzymatic
activity resides in the B chain. The arrow shows a point of
frequent proteolytic cleavage.

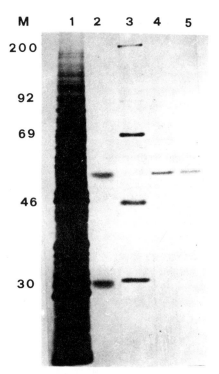

Fig.2 <u>In vitro</u> translation of human kidney polyA+ RNA in a rabbit
reticulocyte cell-free system. 12.5% polyacrylamide gel electrophoresis of the ^{35}S <u>in vitro</u>-synthesized products. Lane
1: total products. Lane 2: immunoprecipitate with polyclonal
anti-urokinase IgG. Lane 3: molecular weights markers. Lanes
4,5; immunoprecipitate with mAb 52C7 (B-chain specific) or
mAb 105 IF10, A-chain specific.

RESULTS
        The first question we have asked is whether the A and B
chains of urokinase are both the products of the same mRNA, in
other words whether they share a common biosynthetic precursor. To
this purpose we have isolated monoclonal antibodies using purified human urokinase as an antigen. Among these antibodies we found
some which recognize antigenic determinants present either on the
A (mAb 105 IF10) or on the B chain (mAb 52C7) (18). We have shown
that both mAbs (and also a polyclonal anti-urokinase antiserum)
immunoprecipitate the urokinase produced by a rabbit reticulo-
cytes cell-free protein synthesizing system (19) directed by total polyA$^+$ RNA extracted from human kidney. The results of such
an experiment are reported in Fig. 2. Both mAbs. precipitate a
protein of about 50,000 daltons (lanes 4 and 5). The polyclonal
antiserum (lane 2), however, also precipitates a band of about
30,000 which is either a cleavage product or a non specific precipitate. These data show that the mature two chains form of urokinase derives from a single-chain prourokinase. This agrees with data
from other laboratories which have shown that an inactive prourokinase is present in the culture medium of several cells (2,
20).Plasmin activates the inactive prourokinase via a proteolytic cleavage forming a 20,000 and a 32,000 daltons chain(2,20).
The results of the in vitro translation experiments (Fig.2), indicate that the urokinase mRNA constitutes between 0.1 and 0.01%
of total polyA$^+$ RNA extracted from human kidneys (18).
        The second question we have tried to answer is the follow-
ing: is regulation of the plasminogen activator actually due to
an increased synthesis of the molecule? Work in other laboratories
indicated that this was the case, since the increase in plasminogen
activator was not observed in actinomycin D or cycloheximide-treated
cells (1, 3-11). However, with the exception of one case (1), a positive proof of an actual increase in prourokinase synthesis had
not yet been obtained. In order to check this point, we screened
several human cell lines for regulation of plasminogen activator.
We chose two lines, A1251 (kidney carcinoma) and A431 (epidermoid
carcinoma (21). As shown in Table I the totality of plasminogen
activator activity in both control and treated cells is accounted
for by urokinase; in fact, all the enzymatic activity is inhibited
by specific anti-urokinase IgG. Table I also shows that tetradeca-
noyl-phorbol-acetate (TPA) increases the level of plasminogen activator in both A1251 and A431 cells,while epidermal growth factor
(EGF) does so in the A431 cells only. TPA induction is
blocked by treating the cells with either actinomycin D or cycloheximide (Ferraiuolo, Stoppelli, Bullock, Lazzaro, Blasi and Pietropaolo, to be published). Treatment of both kinds of cells with dexamethasone decreases the level of plasminogen activator and reduces the effect of both TPA and EGF (Table I). These data indicate

Table 1
Plasminogen activator activity of A1251 and A43
Effect of treatment with TPA, EGF and dexamethasone.

Cell line	Treatment	Plasminogen activator activity*	
		− IgG	+ IgG
A1251	none	150	8
A1251	30 ng/ml TPA	311	9
A1251	50 nM EGF	100	7
A1251	1 mM dexamethasone	35	not done
A1251	same +30 ng/ml TPA	159	not done
A431	none	85	3
A431	30 ng/ml TPA	440	9
A431	50 nM EGF	195	7
A431	1 mM dexamethasone	78	not done
A431	same + 50 nM EGF	56	not done
A431	same + 30 ng/ml TPA	173	not done

*Activity assayed by the ^{125}I plate fibrinolysis test, with (+IgG)
or without (− IgG) of mono-specific anti-urokinase antibodies.

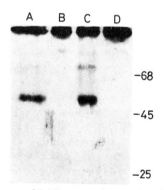

Fig.3 Immunoprecipitation of ^{35}S-labeled urokinase from A1251 medium.
Cells were treated (lanes C,D) or not (lanes A,B) with 30 ng/ml
TPA for 18 h. Lanes A,C:immunoprecipitation with monospecific
antiurokinase rabbit IgG. Lanes B,D: control immunoprecipitation with non immune rabbit IgG. In lanes A and B five times
more radioactive medium was used for immunoprecipitation.

that dexamethasone lowers the expression of the urokinase gene and
hence also reduces the effect of the two inducers. The increase in
plasminogen activator level by EGF in A4341 cells had already been
reported by another group (11); the data of Table I provide the new
information that the activator induced by EGF is in fact urokinase.
The inability of EGF to induce urokinase in A1251 cells may reflect
the lack of EGF-receptors, which on the contrary, are very abundant
in A 431 cells (24). This point deserves further investigation.

    In order to positively show that the increase in urokinase
activity is a consequence of an actual increase of urokinase synthesis, A1251 cells were metabolically labeled with ^{35}-S-L-methio-
nine. Immunoprecipitation with anti-urokinase IgG was carried out
on both the cells lysate and medium. Both in the cells lysate and
in the culture medium a band of about 50,000 daltons is precipita-
ted by antiurokinase antibodies but not by control IgG.The results
for the medium are reported in Fig.3. This band, barely visible in
the control cells (lane A), becomes much more intense when cells
are pre-treated with 30 ng/ml TPA (lane C). An analogous result
has been obtained by immunofluorescence with monoclonal (18) or
polyclonal antibodies (Ferraiuolo et al. to be published). Also
A431 cells, exposed to 30 ng/ml TPA or 50 nM EGF, show the same
effect (data not shown).

    The next question to be answered, then,is whether the ac-
tual increase in urokinase synthesis depends from increase in urokinase mRNA caused by the treatment with TPA or EGF. In order to
study this problem we synthesized an oligonucleotide probe based
on the published aminoacids sequence of urokinase (15-17) in order
to hybridize it to cellular mRNA. The sequence of the synthetic
oligonucleotide is reported in Fig. 4, along with the aminoacids
and mRNA sequence from which it was deduced.

    At first we searched the optimal conditions for hybridiza-
tion and washing of the filters. The ^{32}P-labeled oligonucleotide
was hybridized to RNA samples extracted from human kidney, which
contains urokinase mRNA (18). RNA from non-producing cells (human
fibroblasts and HL60 cells) was used as control.
The hybridization was carried out in 6 x NET for 16 h at 32°.
Filters were washed with 6 x SSPE at different temperatures up to
55°C. The filters washed at temperatures below 37°C gave a positive signal with all tested RNAs. However,when washing was carried
out at 52°, only human kidney and not the control polyA$^+$ RNAs (from
human fibroblasts or HL60 cells),gave a positive signal, nor did
transfer or ribosomal RNA (data not shown). This control was also
complemented by another experiment in which different fractions
of sucrose-gradient-fractionated human kidney polyA$^+$ RNA were assayed for urokinase mRNA by both in vitro translation and dot-blot
hybridization. Human kidney polyA$^+$ RNA was fractionated onto a 15-
23 % sucrose density gradient, for 15 hr at 35,000 rpm, in the
SW41 rotor. Each fraction was translated <u>in vitro</u> in a reticulocyte

```
               15  14  13  12  11  10
HOOC . . . . . PHE TRP PRO GLN ASN GLU . . . . NH2       a.a.sequence


          3'QUC GGU NCC PAC QAA PAG 5'                    RNA sequence


          5' AA CCA NGG QTG PTT QTC 3'                    synthetic DNA
```

Fig.4 Sequence of aminoacids 10-15 of the B chain of human uroki-
      nase (15) and of the derived RNA and synthetic DNA.

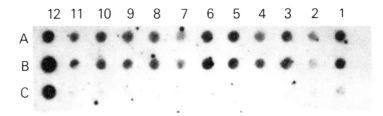

Fig.5 Dot-blot hybridization of ^{32}P-AACCANGGQTGPTTQTC 3'to different
      fractions of human kidney polyA^{+} RNA separated by sucrose den-
      sity (15-23%) gradient centrifugation. In the above sequence,
      N = any nucleotide; Q = a pyrimidine; P = a purine.

Fig.6 Dot-blot hybridization of ^{32}P-AACCANGGQTGPTTQTC 3'to different
      polyA^{+} human RNAs. Each sample was hybridized as such (dots A-E)
      or after alkaline hydrolysis (dots A'-E'). In each case 5 ug of
      RNA was applied. A,A': untreated A1251 cells; B,B': normal
      human fibroblasts; C,C': HL60 cells; E,E': A1251 cells treated
      with 30 ng/ml TPA. Dots D,D' represent 10 ug of A1251 RNA not
      retained by the oligo-dT column, and thus representing mostly
      transfer and ribosomal RNAs.

cell-free system. Urokinase mRNA was located by immunoprecipitation
with monospecific anti-urokinase IgG(to be published). Aliquots of
urokinase positive and negative fractions containing the same amount
of RNA, were blotted onto nitrocellulose filters and hybridized with
the ^{32}P labeled oligonucleotide 5'AACCANGGQTGPTTQTC3'(Fig.4). The
results are shown in Fig. 5. Only the urokinase positive fractions
gave a positive signal. These results thus suggest that the dot-blot
hybridization with the synthetic oligonucleotide provides a specific
assay for urokinase mRNA. We used this assay to compare the amount of
urokinase mRNA in control and TPA-treated A1251 cells (Fig. 6).
PolyA$^+$RNA from untreated A1251 cells (dot A) gave a signal well
above the background of polyA$^+$ RNAs from non producing cells (dots
B,C) or of transfer and ribosomal RNA (dot D). The signal was clearly increased in TPA-treated A1251 cells (dot E). These results show
that the induction of urokinase synthesis by TPA depends on an actual
increase of urokinase mRNA. Whether this is due to an higher rate of
transcription of the urokinase gene or to increased message stability
remains to be investigated. Results similar to those of A1251 cells
have been obtained in A431 cells with either TPA or EGF. In contrast, polyA$^+$ RNA from dexamethasone-treated A1251 or A431 cells,
shows a much weaker hybridization signal (data not shown).

        The oligonucleotide technology outlined above has been used
to screen cDNA libraries prepared from human kidney RNA (from our
laboratory) or from SV40 transformed human fibroblasts RNA (25).
150.000 and 700.000 colonies were screened,respectively, with one
oligonucleotide, and the positive colonies re-tested with three
other oligonucleotides spanning different regions of urokinase.
One clone was positive with all four oligonucleotide probes, as
well as with a nick translated pig urokinase cDNA probe corresponding to the last 50 aminoacids at the COOH terminus (1). Partial
DNA sequencing data have shown that the above clones indeed encode
human urokinase cDNA. The cloning technology and the DNA sequence
of the human urokinase cDNA will be reported elsewhere.

CONCLUSIONS
        In conclusion, we have shown that: 1)the urokinase mRNA
encodes both the A and B chains of the mature enzyme; 2) the amount
of urokinase synthesized is increased by TPA or EGF, and decreased
by dexamethasone; 3) the increase in urokinase synthesis is due to
an increase of the specific mRNA. Finally, 4) we report on the use
of different oligonucleotides for recognizing urokinase mRNA and
for successfully screening cDNA libraries.

ACKNOWLEDGEMENTS
        We are grateful to Drs.Y.Nagamine and E.Reich for providing
the pig urokinase cDNA plasmid; to Dr. S.Aaronson for the A1251
cells, Dr. I.Pastan for the A431 cells; to prof.L. Frati and Prof.
E. Revoltella for pure mouse EGF and EGF antiserum. This work was

supported by grants of the P.F. "Ingegneria Genetica e Basi Moleco-
lari delle Malattie Metaboliche", and "Oncologia" of Consiglio Nazionale delle Ricerche (Italy).

REFERENCES

1. Nagamine,Y., Sudol, M. and Reich,E. 1983 Cell, 32, 1181-1190.
2. Nielsen,L.S., Hansen, J.G., Skriver,L., Wilson,E.L., Kaltoft,K.,
   Zeuthen,J. and Danø, K. 1982 Biochemistry (Wash.), 21, 6410-6415.
3. Unkeless,J.C., Tobia,A.,Ossowski,L., Quigley, J.P., Rifkin,D.B.,
   and Reich,E. 1973 J.Exper.Med. 137, 85-111.
4. Ossowski,L., Unkeless, J.C., Tobia, A., Quigley, J.P., Rifkin,
   D.B., and Reich, E. 1973 J.Exper.Med. 137, 112-126.
5. Rifkin, D.B. 1980 Cold Spring Harbor Symp.Quant.Biol .44,665-668.
6. Strickland, S. and Mahdavi, V. 1978 Cell 15, 393-403.
7. Wigler, M. and Weinstein I.B. 1976 Nature 259 232-233.
8. Weinstein,I.B., Wigler, M., Fisher,P.B., Sisskin, E. and
   Pietropaolo,C. 1978, In Carcinogenesis (T.J.Slaga, A. Sivak and
   R.K.Bontwell, Eds.), Raven Press, New York, N.Y., 2, 313-333.
9. Wilson, E.L., Jacobs, P. and Dowdle,E.B. 1983 Blood 61, 561-567.
10. Lee,L.S. and Weinstein,I.B. 1978 Nature 274, 696-697.
11. Gross, J.L., Krupp, M.N., Rifkin, D.B. and Lane, M.D. 1983 Proc.
    Natl.Acad.Sci.USA 80, 2276-2280.
12. Lin, H.S. and Gordon, S. 1979 J.Exper.Med. 150, 231-245.
13. Rijken,D.C. and Collen,D. 1981 J.Biol.Chem. 256, 7035-7041.
14. Pennica,D., Holmes, W.E., Kohr, W.J., Harkin, R.N., Vehar, G.A.,
    Ward, C.A., Bennett,W.F., Yelverton, E.,Seeburg, P.H., Heyneker,
    H.L., Goeddel, D.V. and Collen, D. 1983 Nature 301, 214-221.
15. Guenzler, W.A., Steffens, G.J., Otting, F.,Buse, G. and Flohe',L.
    1982 Hoppe-Seyler Z.Physiol.Chem. 363, 133-141.
16. Guenzler, W.A., Steffens, G.J., Otting, F., Kim, S.M., Frankus,E.
    and Flohe',L. 1982 Hoppe-Seyler Z.Physiol.Chem. 363, 1155-1165.
17. Steffens,G.J., Guenzler, W.A., Otting, F., Frankus, E. and
    Flohe',L. 1982 Hoppe-Seyler Z. Physiol. Chem. 363, 1043-1058.
18. Salerno, G., Verde, P., Nolli, M.L., Corti, A., Szots, H., Meo,
    T., Johnson, J., Bullock, S., Cassani, G. and Blasi, F. 1984
    Proc.Natl.Acad.Sci.USA, in the press.
19. Pelham, H.R.B. and Jackson, R.J. 1976 Eur. J. Biochem., 67,
    247-256.
20. Wun, T., Ossowski, L. and Reich, E. 1982 J. Biol. Chem. 257,
    7262-7268.
21. Fabricant, R.N., De Larco J.E. and Todaro, G.J. 1977 Proc.Natl.
    Acad. Sci. USA, 74, 565-569.
22. Granelli-Piperno, A. and Reich, E. 1978 J. Exper. Med. 148,
    223-234.
23. Okayama,H. and Berg,P. 1983 Molec.Cell.Biol. 3, 280-289.

CONTROL OF TRANSMITTER RELEASE BY CATECHOLAMINES AND PROSTAGLANDINS

Antonio Capuzzo, Carla Biondi, Pier Giorgio Borasio, Elena Fabbri, Enrica Ferretti and Agostino Trevisani*

Institute of General Physiology, University of Ferrara 44100 Ferrara, Italy

INTRODUCTION

Mammalian superior cervical ganglia (SCG) have long been regarded as a monosynaptic system in which impulse transmission is mediated by the action of acetylcholine (ACh) on nicotinic receptors. In the last two decades, also a muscarinic system[1], a presynaptic feedback mechanism[2] and a postsynaptic inhibitory system[1] have been found to contribute to the regulation of cholinergic transmission.

In this paper we shall summarize the work carried out in our laboratory concerned with the possible role of catecholamines and prostaglandins in decreasing transmitter output through alpha-adrenoceptive sites. The hyperpolarizing response associated with the activation of a postsynaptic receptor[3] will not here be discussed.

RESULTS AND DISCUSSION

Electrical stimulation of preganglionic nerve fibers elicits several types of responses in the principal neurons of SCG. These include a fast excitatory postsynaptic potential (f-EPSP) caused by the activation of nicotinic receptors, a slow excitatory postsynaptic potential (s-EPSP) resulting by ACh interaction with muscarinic receptors, a slow inhibitory postsynaptic potential (s-IPSP)

*Deceased on June 26, 1983

selectively blocked by atropine and by alpha-adrenergic antagonists[4].
To both s-EPSP and s-IPSP has been assigned modulatory functions on
synaptic transmission. A fourth postsynaptic potential has been described, the late slow EPSP. It is not prevented by cholinergic block-
ers and it has been proposed that it is not mediated by ACh[5].

The SIF Cells

Several evidences support the view that intraganglionic inter-
neurons are involved in the genesis of s-IPSP. Indeed, afferent fibers make excitatory synaptic contacts not only with postganglionic
neurons, but also with small intraganglionic ones (the so called
Small Intensely Fluorescent cells).

SIF cells have been demonstrated to contain catecholamine gran-
ules and to make efferent-type synaptic contacts with principal
neurons[6]. These cells may be grouped according to the specific
stored catecholamine (dopamine in rat[7] and rabbit[8], norepinephrine
guinea-pig[9] SCG) or by their structural characteristics. Type I SIF
cells are few in number and possess one or more ramifying processes
among the principal ganglion cells. Type II SIF cells have sparse
and short processes terminating near the adjacent blood vessels.
Rat and guinea-pig SCG exhibit unusual mixed clusters which clearly
contain both types of SIF cells[10].

Catecholamines and Cyclic Nucleotides as Intraganglionic Transmitters

It has been hypothised that ACh released by preganglionic fibers activates not only nicotinic and muscarinic receptors on the
subsynaptic membrane, but also a muscarinic receptor on SIF cells.
This interaction is thought to cause the release of a catecholamine
evoking a s-IPSP as a consequence of its binding to an alpha-recep-
tor on principal neuron membranes[3].

Recently, Greengard and coworkers[11] have suggested that cyclic
AMP mediates the catecholamine-induced slow hyperpolarizing response
and that cyclic GMP mediates the ACh-induced slow depolarizing response. The phosphorylation of a membrane protein by a cAMP-dependent
and a cGMP-dependent protein kinase, respectively, may underlie the
genesis of slow potentials in the mammalian sympathetic ganglia.

However, more recent electrophysiological investigations and experiments on the effects of drugs which affect cyclic nucleotides metabolism do not support this idea. Although theophylline potentiates the s-IPSP, it reduces the amplitude of the s-EPSP whereas papaverine, another phosphodiesterase inhibitor, reduces the amplitude of both s-IPSP and s-EPSP[12]. Little effect of cAMP application has been observed despite the use of different application techniques, different cyclic nucleotide derivatives and wide range of concentrations and time periods[13]. Cyclic AMP has been demonstrated to produce a depolarization, a hyperpolarization or no effect[14].

These observations do not support the hypothesis that the s-PSPs in sympathetic ganglia are mediated by cyclic nucleotides. Indeed, as far as s-IPSP is concerned its physiological role remains obscure, probably related to some unknown metabolic processes.

Presynaptic Mechanism of Catecholamine Action

The depressant action of epinephrine on the frequency of firing discharge in SCG has been originally reported by Marrazzi[15] and subsequently confirmed by many investigators. Yet the controversy about the mechanism of this inhibition has not been fully resolved.

It has been suggested that epinephrine hyperpolarizes the postsynaptic membrane[16] and that it has a dual action in ganglion cells[17]: presynaptic reduction of ACh release and postsynaptic depression of the sensitivity to the neurotransmitter. However, epinephrine fails to hyperpolarize the neuron membrane at the concentration at which it blocks synaptic transmission, suggesting that hyperpolarization is not the primary mechanism of the catecholamine-induced inhibition. Moreover, epinephrine does not appear to reduce the membrane excitability since there is no alteration of threshold depolarization after prolonged perfusion with the catecholamine[18].

Recently it has been suggested that ganglionic transmission is inhibited by dopamine (DA) in rabbit SCG[19] and by norepinephrine (NE) in guinea-pig SCG[20], principally by a reduction of the ACh output from the presynaptic nerve terminals. Studies carried out in our laboratory support the view that, at least in guinea-pig SCG, the depression of synaptic transmission occurs through a mechanism of presynaptic inhibition, and that not only catecholamines but also prostaglandins (PGs) and cyclic nucleotides are involved in this mechanism[21-27].

The Effects of Catecholamines and PGEs on cAMP Levels in Bovine
Ganglia Slices

    The observation that preganglionic stimulation of the nerve fi-
bers in the cervical sympathetic trunk results in the release of
prostaglandins, mainly of the E type[28], prompted us to investigate
the possible functional correlation between catecholamines, prostaglandins and cAMP in sympathetic ganglia.

    It has been observed that when slices of bovine superior cervi-
cal ganglia are incubated in the presence of DA, NE and PGE_1 there
is a significative, dose-dependent increase of cAMP levels. Moreover,
combination of DA and PGE_1 exerts a synergistic effect on cAMP levels
which is not shared by any combination of other PGs with DA or
by other catecholamines with PGE_1 (Fig. 1).

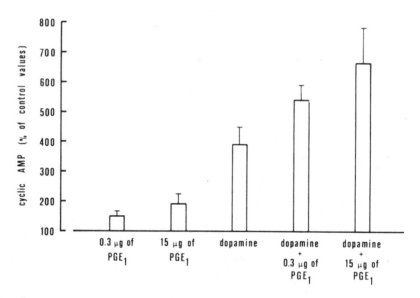

Fig. 1.   The synergistic effect on cAMP concentration by PGE_1 and
          dopamine in bovine SCG slices. All stimulations were sig-
          nificantly different from control. Dopamine was 0.1 mM.
          Basal cAMP concentration was 29.5±3.0 pmol/mg of protein
          ±S.E.M. cAMP was assayed by the method of Brown et al.[29]

The observed synergism may be explained by assuming that PGE_1 acts
on a DA receptor and increases the number of binding sites, or their
affinity, for DA. Neither DA nor PGE_1, alone or in combination, influence cAMP phosphodiesterase suggesting that their effects are due
only to adenylate cyclase stimulation. The latter enzyme of synapto-
somes isolated from the ganglia appears to be responsive to PGE_1 as
the slices but it is poorly stimulated by DA and is not synergisti-
cally modulated by DA in the presence of PGE_1. The different actions
of DA and PGE_1 in the presence of specific antagonists of DA-sensitive adenylate cyclase and other data has been interpreted as indicating the presence of a presynaptic PGE_1-sensitive and a postsynaptic PGE_1-modulated DA-sensitive adenylate cyclase in the bovine superior cervical ganglion.

The Effects of PGEs and Electrical Stimulation on cAMP in SCG of
Different Animal Species

Subsequently, investigations have been undertaken in order to
determine whether a similar effect of PGEs and catecholamines can
be detected in SCG of other animal species. In Fig. 2 the effects
of DA, NE and PGE_1 on cAMP levels in rabbit, guinea-pig, calf and
rat SCG are illustrated. The results indicate that DA is highly ef-

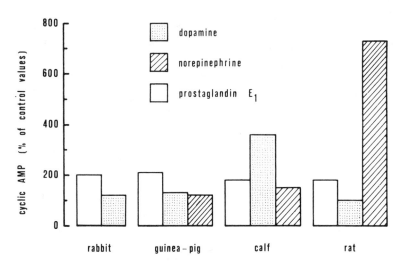

Fig. 2.   The effects of dopamine, norepinephrine and PGE_1 on cAMP
          levels in SCG of different animal species. NE and DA we-
          re 0.1 mM. PGE_1 was tested at 4 μg/ml.

fective only in the bovine SCG, NE only on the rat SCG, whereas the activity of both is either slight or even absent on SCG from rabbit and guinea-pig. On the contrary, PGE_1 is effective on the ganglia of all animal species tested.

These data seem to indicate that the mechanism suggested by Greengard and coworkers[30] cannot be applied to sympathetic ganglia of all animal species.

The Effects of Electrical Stimulation on cAMP and PGEs Levels in Guinea-pig SCG

In guinea-pig preparations, the amount of PGE-like material released in the incubation medium during a 10 min period of electrical stimulation has been determined by radioimmunoassay[31]. PGE levels are nearly doubled (Table 1) and a similar increase of cAMP levels is observed in the ganglia of all animal species tested following electrical stimulation of preganglionic fibers.

As already pointed out, cAMP levels have been suggested as a modulating factor for nervous transmission in peripheral ganglia. Prostaglandins have also been reported to play a role in the modulation of some nervous functions[32] and they could be involved in mammalian SCG as neuromodulators of nervous impulse transmission as a consequence of their effect on cAMP levels.

Table 1. Effect of preganglionic stimulation on cAMP and PGEs levels in SCG of different animal species

Animal	% of control values	
	cAMP	PGEs
Rabbit	204[a]	---
Guinea-pig	197	199
Calf	185[b]	---
Rat	190	---

[a]From Kalix et al.[33]
[b]Experiments carried out on ganglion slices.

In order to exclude the possibility that the observed increase of cAMP could not reflect events directly related to synaptic transmission, we have stimulated ganglia preparations incubated in media containing low Ca^{++} and high Mg^{++} concentrations, a condition known to impair neurotransmitter release from nerve terminals. The data reported in Table 2 show that under these conditions PGE biosynthesis and release appears to be inhibited while cAMP levels too are highly reduced in comparison to those measured in the ganglia superfused with normal solution. The increase of cAMP and PGE levels elicited by electrical stimulation of guinea-pig SCG can be significantly reduced by previous treatment of the preparations with indomethacin (data not shown).

These results suggest the following chain of events in guinea-pig SCG, where synaptic contacts of SIF cells with principal neurons are scarce and catecholamines are unable to reproduce the effect of electrical stimulation on cAMP levels: ACh released by electrical stimulation results in excitation of SIF cells causing NE release, which gives rise to PGE biosynthesis and release. Prostaglandins may, in turn, interact with a presynaptic receptor coupled to adenylate cyclase resulting in cAMP increase with concomitant re-

Table 2. Effect of electrical stimulation on cAMP and PGEs levels in guinea-pig SCG incubated in normal and low Ca^{++}/high Mg^{++} solutions

Experimental conditions	% of control values	
	cAMP	PGEs
(A) Ca^{++} 2.50 mM $\quad Mg^{++}$ 1.15 mM	197 ± 15^a	232 ± 57^a
(B) Ca^{++} 0.50 mM $\quad Mg^{++}$ 12.0 mM	133 ± 17	112 ± 15

Basal levels of cAMP were 10.6 ± 0.3 (A) and 7.3 ± 0.3 (B) pmol/ganglion$\pm$S.E.M. Basal levels of PGEs were 23.3 ± 4.5 (A) and 22.3 ± 1.8 (B) pg/ganglion$\pm$S.E.M. Ganglia were stimulated 10 min at a frequency of 20/sec.
[a]Statistically different with respect to control.

Table 3. Effect of electrical stimulation on cAMP and PGEs
          levels in guinea-pig SCG in the presence of cholin-
          ergic muscarinic and adrenergic antagonists

Experimental conditions	% of control values	
	cAMP	PGEs
Stimulated ganglia	202 ± 9^a	222 ± 39^a
Stimulated ganglia + atropine 10^{-6}M	149 ± 22	95 ± 6
Stimulated ganglia + phentolamine 10^{-5}M	148 ± 27	100 ± 24
Stimulated ganglia + propranolol 10^{-5}M	205 ± 10^a	187 ± 9^a

Basal level of cAMP was 7.6 ± 1.1 pmol/ganglion$\pm$S.E.M. Basal
level of PGEs was 26.5 ± 2.0 pg/ganglion$\pm$S.E.M. Conditions
as in Table 2.
[a]Statistically different with respect to control.

duction of the release of ACh. In order to clarify this mechanism,
selective agents which block muscarinic and alpha-adrenergic receptors have been tested.

As shown in Table 3, atropine and phentolamine markedly reduce
PGEs and cAMP increase. In contrast, propranolol does not appreciably affect either PGEs release or cAMP rise suggesting that PGEs
biosynthesis and release induced by electrical stimulation of preganglionic fibers requires both a cholinergic muscarinic and an
alpha-adrenergic receptor.

The Effects of PGEs and NE on Ganglionic Synaptic Transmission

The presynaptic site of action of PGEs has been conclusively
demonstrated by the addition of low concentrations of PGE_1 to the
perfusion bath of guinea-pig SCG (Fig. 3). This causes a remarkable
impairment of synaptic transmission; after wash out of the bath
with plain Krebs solution a complete recovery of the neuron response
is detected. Indeed, the rapid and reversible block of synaptic

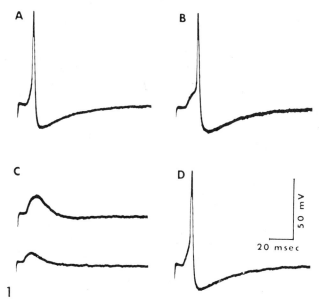

Fig. 3.   Electrical activity recorded from a single guinea-pig SCG
          neuron in the absence (A) and in the presence (B-C) of
          10^{-7}M PGE_1. (D) Recovery after 15 min in normal Krebs so-
          lution. Stimulation frequency: 1 Hz.

transmission by PGE_1 takes place without affecting the resting membrane potential or the membrane resistance. Furthermore a reduction
in the number of elementary quanta liberated is observed while, on
the contrary, quantal size appears to be unchanged.

The possibility that the inhibitory action of PGE_1 may be mediated by NE has been excluded by blocking alpha-adrenoceptors with
phentolamine added to the superfusing medium. In this condition, no
difference can be detected in the ability of PGE_1 to impair synaptic
transmission in the ganglia.

NE release by the SIF cells has been shown to be a prerequisite
condition for PGEs to be liberated from the ganglia following
electrical stimulation of the nerve fibers in the cervical sympathetic trunk. At the same time, NE can exert a direct effect on presynaptic structures[20]. Thus the control of ACh output appears to be
brought about by PGEs and NE through independent interactions with
specific presynaptic receptors, since alpha-adrenergic block does
not influence the inhibitory activity of exogenous PGEs.

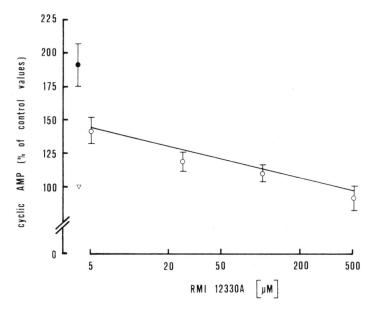

Fig. 4.   Inhibition by RMI 12330A of PGE_2-induced cAMP accumulation
           in guinea-pig SCG. Basal level was 40.8 ± 2.4 pmol/mg pro-
           tein$\pm$S.E.M. $\triangledown$,control; $\bullet$, 10 μM PGE_2; °, 10 μM PGE_2 + RMI
           12330A.

This mechanism has been further confirmed by monitoring intracellular potentials from impaled neurons exposed to PGEs and NE, alone
or in combination with RMI 12330A, a potent inhibitor of adenylate
cyclase activity[27] whose effect on PGE_2-induced cAMP accumulation
is shown in Fig. 4.

    Both NE and PGE_2 inhibit the firing discharge of guinea-pig
SCG neurons electrically stimulated. SCG superfusion with RMI 12330A
does not affect the action potentials elicited by preganglionic stimulation nor alter the membrane resting potential and input resistance
of principal neurons. When RMI 12330A is tested together with
PGE_2 the inhibitory effect of the prostaglandin is prevented. On
the other hand, RMI 12330A in combination with NE is unable to affect the inhibitory action of the catecholamine confirming that NE
and PGE_2 act through different pathways (Fig. 5).

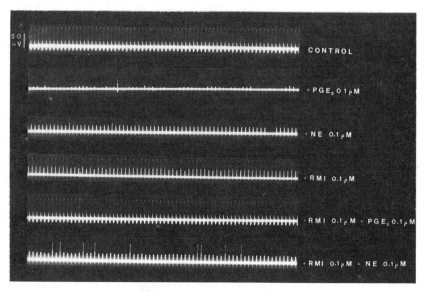

Fig. 5.   Firing discharge of guinea-pig SCG neurons stimulated at
1 Hz and superfused with 0.1 μM PGE$_2$, 0.1 μM NE, 0.1 μM
RMI 12330A, alone or in combinations.

Electrically Evoked Release of ^{3}H-ACh from Guinea-pig SCG

In order to differentiate the two hypothesized mechanisms, experiments have been performed investigating the effect of different
substances on the electrically evoked overflow of tritium from guinea-pig SCG preincubated with ^{3}H-choline. It is assumed that preincubation of nervous tissue in the presence of low concentrations of
^{3}H-choline leads to selective uptake of ACh precursor into cholinergic neurons[34]. The stimulation evoked tritium outflow from SCG preincubated with ^{3}H-choline would reflect labeled ACh release as shown
for other tissues[35,36].

Application of forskolin, a diterpene derivative which acti-
tes adenylate cyclase in membranes from a variety of tissues[37], including CNS, significantly reduces (21% inhibition) neurotransmitter release in guinea-pig SCG. On the other hand, forskolin dose-
dependently stimulates adenylate cyclase activity in synaptosomal
preparations reaching a 350% stimulation at 100 μM concentration
and elevates cAMP levels in intact SCG (data not shown).

^{3}H-ACh release has been then measured in the presence of spe-

Table 4. Effects of alpha-adrenergic receptor agonists, alone
          or in the presence of RMI 12330A, on electrically
          evoked tritium overflow from guinea-pig SCG preincu-
          bated with ^{3}H-choline

Experimental conditions	Stimulation evoked overflow of tritium (S_2/S_1)
Control	0.96 ± 0.05
Clonidine 1 µM	0.43 ± 0.04^a
Clonidine 1 µM + RMI 12330A 1µM	$0.43 \pm 0.08^{a,b}$
Phenylephrine 10 µM	0.72 ± 0.06^a
Phenylephrine 10 µM + RMI 12330A 1 µM	$0.93 \pm 0.01^{c,d}$

The stimulated evoked tritium overflow is expressed as the ratio of the tritium overflow (as % of ganglion content) evoked
in two 5 min stimulation periods (S_2/S_1). Drugs were added 15
min before S_2.
[a]Significantly different with respect to control.
[b]Not significantly different with respect to clonidine.
[c]Not significantly different with respect to control.
[d]Significantly different with respect to phenylephrine.

cific alpha-adrenergic agonists. As shown in Table 4, alpha$_1$-adren-
ergic agonist phenylephrine significantly reduces tritium outflow
to 75% of control while alpha$_2$-adrenergic agonist clonidine is more
efficient causing a 55% inhibition. When both agonists are tested
in the presence of the adenylate cyclase inhibitor RMI 12330A, the
effect of phenylephrine but not that of clonidine is prevented, suggesting that alpha$_1$-agonist action may involve a cAMP-dependent step.

CONCLUSIONS

    The data summarized in this article suggest that synaptic
transmission modulation in guinea-pig SCG could partly take place
as illustrated in Fig. 6. ACh released from presynaptic nerve terminals interacts with a postsynaptic nicotinic receptor giving rise

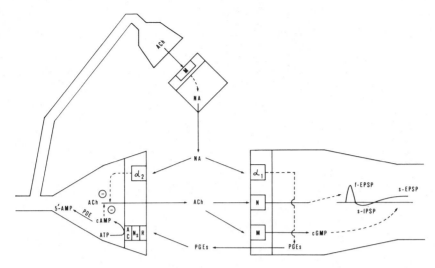

Fig. 6.   A simplified schematic diagram of the principal synaptic
          connections in the guinea-pig SCG. ACh and NE interactions
          with their specific receptors, the proposed involvement of
          PGEs and cyclic nucleotides, the electrical signals evoked
          in postganglionic structures are indicated.

to f-EPSP; in addition, the transmitter induces NE release through a
cholinergic muscarinic receptor on the SIF cells. The catecholamine,
in turn, can depress further release of ACh directly, through an
interaction with a presynaptic receptor, probably of the $alpha_2$
type, and indirectly through an interaction with a receptor, probably of $alpha_1$ type, on the principal neuron, resulting in PGEs synthesis and release. PGEs could stimulate adenylate cyclase activity
and induce cAMP accumulation in presynaptic structures.

The change in nerve terminal function produced by both NE and
cAMP which provokes a reduction of ACh release remains to be identified. It has been suggested that presynaptic inhibition involves
a modulation of the calcium current[38]. Whether the altered calcium
current results directly from an action on the calcium channel or
indirectly from an action on the opposing potassium current has not
yet been determined.

REFERENCES

1. R. M. Eccles and B. Libet, Origin and blockade of the synaptic responses of curarized sympathetic ganglia, J. Physiol. (London) 157:484 (1961).

2. D. D. Christ and S. Nishi, Site of adrenaline blockade in the superior cervical ganglion of the rabbit, J. Physiol. (London) 213:107 (1971).

3. B. Libet, Generation of slow inhibitory and excitatory postsynaptic potentials, Fed. Proc. 29:1945 (1970).

4. B. Libet, Slow postsynaptic actions in ganglionic functions, in:"Integrative functions of the autonomic nervous system", C. McC. Brooks, K. Koizumi and A. Sato, eds., Elsevier/North Holland Biomedical Press, Amsterdam (1979).

5. S. Nishi and K. Koketsu, Early and late after-discharges of amphibian sympathetic ganglion cells, J. Neurophysiol. 31: 109 (1968).

6. O. Eranko and M. Harkonen, Histochemical demonstration of fluorogenic amines in the cytoplasm of sympathetic ganglion cells of the rat, Acta Physiol. Scand. 58:285 (1963).

7. A. Bjorklund, L. Cegrell, B. Falck, M. Ritzen and E. Rosengren, Dopamine-containing cells in sympathetic ganglia, Acta Physiol. Scand. 78:334 (1970).

8. B. Libet and C. H. Owman, Concomitant changes in formaldehyde-induced fluorescence of dopamine interneurones and in slow inhibitory postsynaptic potentials of rabbit superior cervical ganglion induced by stimulation of preganglionic nerve or by a muscarinic agent, J. Physiol. (London) 237:635 (1974).

9. L.-G. Elfvin, T. Hokfelt and M. Goldstein, Fluorescence microscopical, immuno-histochemical and ultrastructural studies on sympathetic ganglia of the guinea-pig, with special reference to the SIF cells and their catecholamine content, J. Ultrastruct. Res. 51:377 (1975).

10. A. C. Black, Jr., J. K. Wamsley, D. Sandquist and T. H. Williams, The guinea pig further demonstrates that rodent superior cervical ganglia differ from those of other species, Exp. Neurol. 77:314 (1982).

11. P. Greengard, Possible role for cyclic nucleotides and phosphorylated membrane proteins in postsynaptic actions of neurotransmitters, Nature 260:101 (1976).

12. N. A. Busis, F. F. Weight and P. A. Smith, Synaptic potentials

in sympathetic ganglia: Are they mediated by cyclic nucleotides?, Science 200:1079 (1978).

13. F. F. Weight, P. A. Smith and J. A. Schulman, Postsynaptic potential generation appears independent of synaptic elevation
of cyclic nucleotides in sympathetic neurons, Brain Res.
158:197 (1978).

14. J. P. Gallagher and P. Shinnick-Gallagher, Cyclic nucleotides
injected intracellularly into rat superior cervical ganglion
cells, Science 198:851 (1977).

15. A. S. Marrazzi, Adrenergic inhibition at sympathetic synapses,
Am. J. Physiol. 127:738 (1939).

16. A. Lundberg, Adrenaline and transmission in the sympathetic
ganglion of the cat, Acta Physiol. Scand. 26:251 (1952).

17. W. D. M. Paton and J. W. Thompson, The mechanism of action of
adrenaline on the superior cervical ganglion of the cat,
Int. Physiol. Congr. 19:664 (1953).

18. S. Nishi, The catecholamine-mediated inhibition in ganglionic
transmission, in:"Integrative functions of the autonomic
nervous system", C. McC. Brooks, K. Koizumi and A. Sato,
eds., Elsevier/North Holland Biomedical Press, Amsterdam
(1979).

19. N. Dun and S. Nishi, Effects of dopamine on the superior cervical ganglion of the rabbit, J. Physiol. (London) 239:155
(1974).

20. N. Dun and A. G. Karczmar, The presynaptic site of action of
norepinephrine in the superior cervical ganglion of guinea
pig, J. Pharmacol. Exp. Ther. 200:328 (1977).

21. V. Tomasi, C. Biondi, A. Trevisani, M. Martini and V. Perri,
Modulation of cyclic AMP levels in the bovine superior cervical ganglion by prostaglandin E_1 and dopamine, J. Neurochem. 28:1289 (1977).

22. V. Tomasi, A. Trevisani, C. Biondi, A. Capuzzo and V. Perri,
The role of prostaglandin E_1 as an intercellular regulator
or modulator of adenylate cyclases, Biochem. Soc. Trans.
5:520 (1977).

23. O. Belluzzi, C. Biondi, P. G. Borasio, A. Capuzzo, M. E. Ferretti, A. Trevisani and V. Perri, Influence of prostaglandins of E type on synaptic transmission of the guinea-pig
superior cervical ganglion, in:"Adv. Physiol. Sci. Vol. 4.
Physiology of Excitable Membranes", J. Salanki, ed., Pergamon Press, New York (1981).

24. V. Perri, O. Belluzzi, C. Biondi, P. G. Borasio, A. Capuzzo,

M. E. Ferretti and A. Trevisani, PGE_1-induced cAMP biosynthesis in the superior cervical ganglion of different animal species, in:"Cholinergic Mechanisms: Phylogenetic Aspects, Central and Peripheral Synapses, and Clinical Significance", G. Pepeu and H. Ladinsky, eds., Plenum Press, New York (1981).

25. A. Trevisani, C. Biondi, O. Belluzzi, P. G. Borasio, A. Capuzzo, M. E. Ferretti and V. Perri, Evidence for increased release of prostaglandins of E-type in response to orthodromic stimulation in the guinea-pig superior cervical ganglion, Brain Res. 236:375 (1982).

26. O. Belluzzi, C. Biondi, P. G. Borasio, A. Capuzzo, M. E. Ferretti, A. Trevisani and V. Perri, Electrophysiological evidence for a PGE-mediated presynaptic control of acetylcholine output in the guinea-pig superior cervical ganglion, Brain Res. 236:383 (1982).

27. P. G. Borasio, M. E. Ferretti, C. Biondi and A. Trevisani, cAMP-dependent and cAMP-independent modulation of synaptic transmission in guinea-pig superior cervical ganglion, Neurosci. Lett. 32:197 (1982).

28. H. A. Davis, E. W. Horton, K. B. Jones and J. P. Quilliam, Identification of prostaglandins in prevertebral venous blood after preganglionic stimulation of the cat superior cervical ganglion, Br. J. Pharmac. 42:569 (1971).

29. B. L. Brown, R. P. Ekins and J. D. M. Albano, Saturation assay for cyclic AMP using endogenous binding protein, in: "Advances in cyclic nucleotides research", Vol. 2, P. Greengard, R. Paoletti and G. A. Robison, eds., Raven Press, New York (1972).

30. P. Greengard, D. A. McAfee and J. W. Kebabian, On the mechanism of action of cyclic AMP and its role in synaptic transmission, in:"Advances in cyclic nucleotide research", Vol. 1, P. Greengard, R. Paoletti and G. A. Robison, eds., Raven Press, New York (1972).

31. G. Bartolini, C. Meringolo, M. Orlandi and V. Tomasi, Biosynthesis of prostaglandins in parenchymal and nonparenchymal rat liver cells, Biochim. Biophys. Acta 530:325 (1978).

32. P. Hedqvist, Prostaglandin action on transmitter release at adrenergic neuroeffector junctions, in:"Advances in Prostaglandin and Thromboxane Research", Vol. 1, B. Samuelsson and R. Paoletti, eds., Raven Press, New York (1976).

33. P. Kalix, D. A. McAfee, M. Schorderet and P. Greengard, Pharmacological analysis of synaptically mediated increase in

cyclic AMP in rabbit superior cervical ganglion, J. Pharmacol. Exp. Ther. 188:676 (1974).

34. R. S. Jope, High affinity choline transport and acetylCoA production in brain and their roles in the regulation of acetylcholine synthesis, Brain Res. Rev. 1:313 (1979).

35. A. H. Mulder, H. I. Yamamura, M. J. Kuhar and S. H. Snyder, Release of acetylcholine from hippocampal slices by potassium depolarization: dependence on high affinity choline uptake, Brain Res. 70:372 (1974).

36. G. Hertting, A. Zumstein, R. Jackkisch, I. Hoffman and K. Starke, Modulation by endogenous dopamine of the release of acetylcholine in the caudate nucleus of the rabbit, Naunyn-Schmied. Arch. Pharmacol. 315:111 (1980).

37. J. W. Daly, W. Padgett and K. B. Seamon, Activation of cyclic AMP-generating systems in brain membranes and slices by the diterpene forskolin: augmentation of receptor-mediated responses, J. Neurochem. 38:532 (1982).

38. M. Klein, E. Shapiro and E.R. Kandel, Synaptic plasticity and the modulation of the Ca^{2+} current, J. Exp. Biol. 89:117 (1980).

GANGLIOSIDE INVOLVEMENT IN TROPHIC INTERACTIONS

Daniela Benvegnù, Donatella Presti, Laura Facci, Roberto
Dal Toso, Alberta Leon and Gino Toffano

Fidia Research Laboratories, Department of Biochemistry
Via Ponte della Fabbrica 3/A, 35031 Abano Terme, Italy

INTRODUCTION

Although the severance of neuronal projections in the mammalian Central Nervous System (CNS), in contrast to Peripheral
Nervous System (PNS), is usually followed by only a limited regrowth of axons, it has already been shown that many neurons in
the adult mammalian CNS are inherently capable of reorganizing or
reforming their synaptic connections in response to lesions or
other perturbations (Björklund and Stenevi, 1979; Cotman et al.,
1981). The understanding of the molecular processes underlying the
recovery of brain function is still to be solved. The advent of
tissue culture techniques, grafting and transplantation procedures
is now providing new insights and perspectives to the problem of
CNS regeneration (Björklund et al., 1980; Bignami et al., 1981;
Benfey and Aguayo, 1982; Nieto-Sampedro et al., 1983). It is
widely accepted that the regrowth of axons and establishment of
functional connections in the adult CNS, in a manner similar to
the ability of axons to grow and connect specific target areas
during development, is largely controlled by its humoral and
cellular microenvironment which provides a suitable terrain with
appropriate directional cues over which to grow (Varon and Adler,
1980; Varon and Adler, 1981; Aguayo et al., 1982; Thoenen et al.,
1982; Varon et al., 1983). For successful regeneration, an adequate stimulus must also be conveyed retrogradely from the site of
injury to the parent nerve cell body. The stimulus presumably
induces a series of biochemical modifications the outcome of which
ultimately determines the fate of the injured neuron: survival
with or without regeneration or death (Willard and Skene, 1982).
The irreversibly damaged neurons in the brain cannot be replaced.
Nevertheless their substitution by alternative pathways, e.g.

reactive synaptogenesis of undamaged axons, as well as regrowth of damaged axons of surviving neurons has been shown to occur (Cotman et al., 1981).

Starting from these premises, the identification and isolation of the proposed extrinsic factors have to be considered of great importance not only for understanding the mechanisms controlling neuronal plasticity but also for deriving new pharmacological strategies aiming to improve the recovery of brain function after injuries.

GANGLIOSIDES AND NEUROTROPHIC INTERACTIONS

Since the discovery that gangliosides exist at high concentrations in the brain (Klenk, 1942), there has been much speculation concerning their role in CNS. Because of the diversity of ganglioside structure, their preferential orientation in the outer leaflet of membrane bilayer with their oligosaccharide portion extending into the external environment, gangliosides have been assumed to play a major role in modulation of receptor function and biotransduction of membrane-mediated information (Ledeen, 1978; Sharom and Grant, 1978; Yamakawa and Nagai, 1978; Hakomori, 1981).

Studies employing lectin labelling of sprouting neurons have recently suggested that gangliosides may be the predominant form of glycoconjugate in the growing neurite (Pfenninger and Maylie-Pfenninger, 1978) and consequently may contribute significantly to the regulation of neurite outgrowth of both developing and mature neurons (Purpura and Baker, 1978; Dreyfus et al., 1980; Willinger and Schachner, 1980). Thus it has been suggested that gangliosides may have a role as mediator molecules of neurotrophic interactions regulating neurite outgrowth (Morgan and Seifert, 1979).

Approaches for inspecting such a prospective role of gangliosides presently involve the study of the effects of exogenous gangliosides in neuronal tissue culture systems or in animals after lesioning the brain. Along this line, it has been reported that gangliosides and, in particular, GM_1 monosialoganglioside (nomenclature according to Svennerholm, 1963) in vitro markedly stimulates neurite outgrowth of various clonal and primary neuronal cells in culture (Morgan and Seifert, 1979; Dimpfel et al., 1981; Hauw et al., 1981; Roisen et al., 1981; Leon et al., 1982; Dreyfus et al., 1983; Facci et al., 1983; Ferrari et al., 1983) and in vivo is successful in improving CNS nerve regeneration. Chronic GM_1 administration facilitates the functional dopaminergic reinnervation of the striatum and the survival of dopaminergic cell bodies in the substantia nigra in rats with unilateral hemitransection (Agnati et al., 1983; Toffano et al., 1983). Similarly, GM_1 treatment has been shown to intensify the recovery of cholinergic nerve terminals and of behavioural performances in the partially denervated hippocampus following entorhinal cortex (Karpiak, 1983a) or medio-ventral septal lesions (Oderfeld-Nowak et al., 1983). Moreover in neonate rats GM_1 administration has been demonstrated to enhance functional neuronal maturation (Karpiak, 1983b), while antibodies to GM_1 cause subtle behavioural dysfunctions accompanied by morphological abnormalities of dendritic arborization (Kasarskis et al., 1981).

GM_1 GANGLIOSIDE AND NERVE GROWTH FACTOR ACTIVITY

To date, Nerve Growth Factor (NGF) is the only macromolecular agent directed to neurons that has been fully purified and characterized. NGF is known to be essential for the survival, neurite outgrowth, proper development, and presumably also maintenance and regeneration of its target sympathetic and sensory ganglionic neurons both in vitro and in vivo (Thoenen and Barde, 1980; Varon and Adler, 1980; Harper and Thoenen; 1981; Varon and Adler, 1981; Levi-Montalcini, 1982). Indeed, in newborn rats and mice, the interruption of axonal transport of NGF results in degeneration of a majority of the sympathetic neurons as does the neutralization of endogenous NGF by anti-NGF antibodies. This effect can be prevented by the administration of exogenous NGF. Similarly, dorsal root sensory ganglionic (DRG) cells, known to require NGF for survival and acquisition of differentiated properties only during a limited period of embryonic development, are now known to require NGF even after birth if they are deprived of their ganglionic non-neural partners. Consequently, NGF is currently viewed as a prototype for other hypothetical, macromolecular factors directed to the control of neuronal maintenance, growth and regeneration.

A molecular approach to the problem of possible ganglioside mediation of neuronotrophic activity may as such be exemplified, as described below, by the study of monosialoganglioside potentiation of NGF-induced neurite outgrowth from fetal DRG cells in culture. It is recognized that the study of the NGF effects on their neurite production in vitro concerns regeneration rather than de novo induction and hence provides unique opportunities for the study of the fine interplay occurring between the intrinsic neuronal cellular activities and the extrinsic environmental factors regulating regrowth of neurite extensions (Varon and Adler, 1980).

In brief, dorsal root ganglia obtained from 8 day chick embryos were plated on various substrata both as explant cultures or as dissociated cells. Addition of GM_1 at concentrations ranging from 10^{-7} to 10^{-8} M to culture medium of DRG explants significantly enhances the NGF-induced neurite outgrowth and neurite complexity (Fig. 1.). Quantitative morphometric analysis indicated that

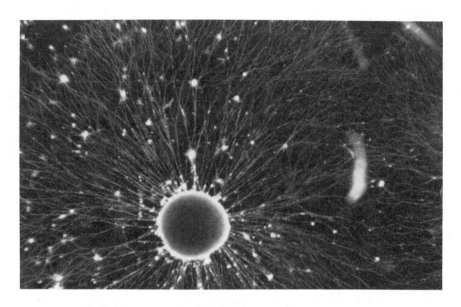

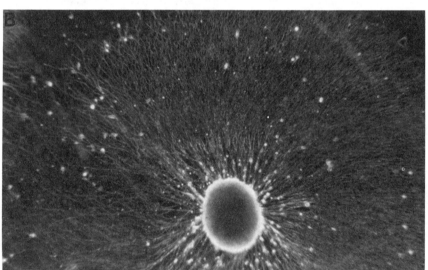

Fig. 1.   Phase contrast micrographs of GM_1 effects on NGF-induced
           neurite outgrowth of DRG explants after 2 days in cul-
           ture: control (A), GM_1 treated (B). Dorsal root ganglia
           were dissected from 8-9 day old chick embryos. Culture
           medium consisted of DMEM + 199 (3:1/vol), 10% fetal calf
           serum, and 10 ng/ml NGF without (A) or with 10^{-7} M GM_1
           (B). Polyornithine (100 μg/35 mm Falcon dishes) pre-
           treated with heart conditioned medium was utilized as
           substratum (Adler et al., 1981).

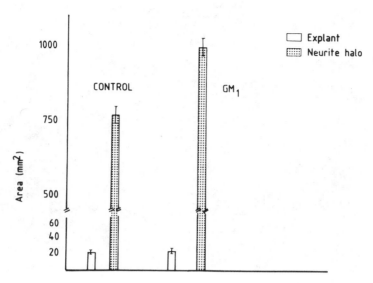

Fig. 2.   Effect of GM$_1$ on NGF-induced neurite outgrowth of DRG
           explants: measurement of area occupied by neurite exten-
           sions. Computer analysis of glutaraldehyde-fixed DRG
           explants after 2 days in culture. Same culture condi-
           tions as in Fig. 1. Values are mean of triplicate analy-
           sis (4 explants/dish) of these experiments ± S.E.
           Control = 10 ng/ml NGF; GM$_1$ treated = 10 ng/ml NGF +
           10^{-7} M GM$_1$.

the effect is associated with a significant increase in total
surface area occupied by the neuritic extensions (Fig. 2), the
entity of which is highly dependent on the NGF and GM$_1$ concentration present in the culture medium. Such an effect was seen to
occur independently of the type of cell substratum utilized (col-
lagen, polyornithine, or polyornithine pretreated with heart
conditioned medium) and on the presence or absence of serum in
culture medium. No effect of GM$_1$ was observed in absence of NGF
indicating no trophic activity per se of the GM$_1$ molecule in this
model. Addition of GM$_1$ to 90-95% pure dissociated fetal DRG neuronal cells, obtained following trypsinization and pre-plating techniques (Barde et al., 1980), similarly enhanced the NGF-induced
neurite outgrowth and neurite complexity thereby excluding an
effect of GM$_1$ on non-neuronal cells present in explant cultures.
This effect was associated with an increase in apparent neurite

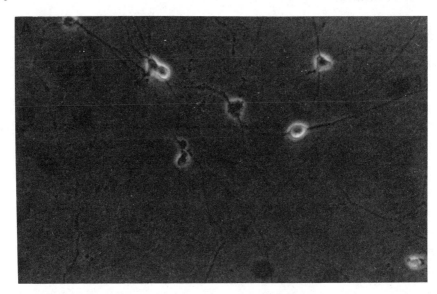

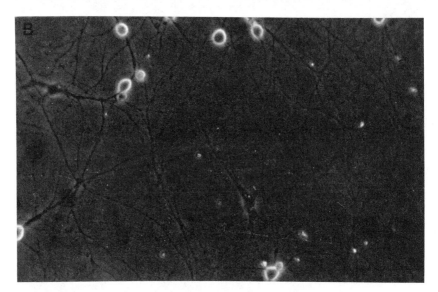

Fig. 3.   Phase contrast micrographs of GM_1 effects on NGF-induced
            neurite outgrowth of DRG dissociated neuronal cells
            after 2 days following withdrawal of NGF. Cell culture
            substratum and medium as in Fig. 1. See text for further
            details: A) control, B) GM_1-treated.

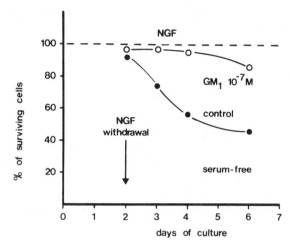

Fig. 4.   Effect of GM$_1$ on the survival of dissociated embryonic
            chick DRG cells following NGF withdrawal. Same culture
            conditions, NGF and GM$_1$ treatment as described in text.
            Surviving cells were evaluated following glutaraldehyde
            fixation as previously described (Barde et al., 1980).

outgrowth (Fig. 3) and neuronal cell survival following NGF withdrawal (Fig. 4) (Green, 1977). In this case, on day 0, cells were
plated on polyornithine substratum pretreated with heart conditioned medium in the presence of 10% fetal calf serum and NGF (10
ng/ml) (control) or NGF + GM$_1$ 10^{-7} M (GM$_1$-treated). After 48
hours, medium was substituted and replaced with serum-free N$_2$
medium without NGF in absence (Fig. 3A) or presence (Fig. 3B) of
GM$_1$. GM$_1$ (10^{-7} M) was re-added to cultures previously exposed to
the ganglioside. The decreased NGF dependancy for survival and
maintenance of neurite extensions of the GM$_1$-treated cells is
indicative of GM$_1$-induced enhanced maturation when initially
cultured in presence of NGF (Green, 1977).

    Analogously GM$_1$ has been reported to potentiate NGF-induced
neurite outgrowth of pheochromocytoma PC12 cells (Ferrari et al.,
1983) whereas addition of ganglioside mixtures (Hauw et al., 1981;
Roisen et al., 1981) or GM$_1$ antibodies have been shown to respectively stimulate or block NGF-induced sprouting from DRG explants
(Schwartz and Spirman, 1982).

CONCLUSION

The aim of this chapter was to provide some perspective evidences highlighting the possibility of improving CNS neuronal repair processes via adequate pharmacological strategies. This may be obtained either by minimizing the effect of deleterious environmental situations or by increasing the potency of existing growth-promoting substances present at the site of lesion (Varon and Adler, 1981; Manthorpe et al., 1983; Nieto-Sampedro et al., 1983; Varon et al., 1983).

Gangliosides, neuronal cell surface components, have been suggested to behave as mediator molecules in trophic interactions. Recent pharmacological attempts utilizing monosialoganglioside GM_1 have been successful in improving functional CNS regeneration. The mechanisms involved are however still obscure. The addition of GM_1 to fetal dorsal root ganglionic neurons in culture amplifies the NGF-induced neurite outgrowth, while GM_1 antiserum and anti-GM_1 antibodies inhibit neurite extension both in goldfish retinal and chick embryo DRG explants in response to a brain factor extract or NGF, respectively (Spirman et al., 1981; Schwartz and Spirman, 1982). Such results are consistent with the view that the presence of GM_1 on the neuronal cell surface, either endogenous or stably inserted exogenous molecules, plays a role in determining the properties of the plasma membrane particularly with respect to the capacity of the latter to transfer and modulate signals from the extra to the intracellular spaces. Study of the underlying molecular mechanisms, utilizing the prototype neuronotrophic factor NGF, may prove to be relevant for the comprehension of the facilitatory effect of chronic GM_1 administration on CNS neuronal regeneration in vivo.

REFERENCES

Adler, R., Manthorpe, M., Skaper, S. D., and Varon, S., 1981, Polyornithine-attached neurite-promoting factors (PNPFs). Culture sources and responsive neurons, Brain Res., 206:129.

Agnati, L. F., Fuxe, K., Calzà, L., Benfenati, F., Cavicchioli, L., Toffano, G., and Goldstein, M., 1983, Gangliosides increase the survival of lesioned nigral dopamine neurons and favour the recovery of dopaminergic synaptic function in striatum of rats by collateral sprouting, Acta Physiol. Scand., in press.

Aguayo, A. J., Richardson, P. M., David, S., Benfey, M., 1982, Transplantation of neurons and sheath cells – A tool for the study of regeneration, in: "Repair and Regeneration of the Nervous System," J. G. Nicholls, ed., Springer-Verlag, Berlin.

Barde, Y.-A., Edgar, H., and Thoenen, H., 1980, Sensory neurons in culture: changing requirements for survival factors during embryonic development, Proc. Natl. Acad. Sci. USA, 77:1199.

Benfey, M., and Aguayo, A. S., 1982, Extensive elongation of axons from rat brain into peripheral nerve grafts, Nature, 296: 150.

Bignami, A., Nguyen, H. C., and Dahl, D., 1981, Centrally cut axons regenerate in peripheral nerve implanted in murine brain: immunofluorescence study with neurofilament and GFA antisera, in: "Post-traumatic Peripheral Nerve Regeneration: Experimental Basis and Clinical Applications," A. Gorio, M. Millesi and S. Mingrino, eds., Raven Press, New York.

Björklund, A., Dunnett, S. B., Stenevi, U., Lewis, M. E., and Iversen, L., 1980, Reinnervation of the denervated striatum by substantia nigra transplants: functional consequences as revealed by pharmacological and sensory motor testing, Brain Res., 199:307.

Björklund, A., and Stenevi, U., 1979, Regeneration of monoaminergic and cholinergic neurons in the mammalian central nervous system, Physiol. Rev., 59:62.

Cotman, C. W., Nieto-Sampedro, M., and Harris, E. W., 1981, Synapse replacement in the nervous system of adult vertebrates, Physiol. Rev., 61:684.

Dimpfel, W., Moller, W., and Mengs, U., 1981, Ganglioside-induced formation in cultured neuroblastoma cells, in: "Gangliosides in Neurological and Neuromuscular Function, Development and Repair," M. M. Rapport and A. Gorio, eds., Raven Press, New York.

Dreyfus, H., Gorio, A., Ferret, B., Hoffman, D., Dainous, F., Freysz, L., and Massarelli, R., 1983, Morphological and neurochemical changes induced by exogenous gangliosides in neuronal membranes, in: abstract book "The Cell Biology of Neuronal Plasticity,", 6–10 June, Villasimius.

Dreyfus, H., Jourdan, J. C., Harth, S., and Mandel, P., 1980, Gangliosides in cultured neurons, Neuroscience, 5:1647.

Facci, L., Leon, A., Toffano, G., Sonnino, S., Ghidoni, R., and Tettamanti, G., 1983, Promotion of neuritogenesis in mouse neuroblastoma cells by exogenous gangliosides. Relationship between the effect and cell association of ganglioside GM_1, J. Neurochem., in press.

Ferrari, G., Fabris, M., and Gorio, A., 1983, Gangliosides enhance neurite outgrowth in PC12 cells, Develop. Brain. Res., 8:215.

Green, L. A., 1977, Quantitative in vitro studies of the nerve growth factor (NGF) requirement of neurons. II. Sensory neurons, Develop. Biol., 58:106.

Hakomori, S. I., 1981, Glycosphingolipids in cellular interaction, differentiation and oncogenesis, Ann. Rev. Biochem., 50:733.

Harper, G. P., and Thoenen, H., 1981, Target cells, biological effects, and mechanism of action of nerve growth factor and its antibodies, Ann. Rev. Pharmacol. Toxicol., 21:205.

Hauw, J. J., Fenelon, S., Boutry, J. M., Nagai, Y., and Escourol-
le, R., 1981, Effects of brain gangliosides on neurite growth
in guinea pig spinal ganglia tissue cultures and on fibroblast cell cultures, in: "Gangliosides in Neurological and
Neuromuscular Function, Development and Repair", M. M. Rapport and A. Gorio, eds., Raven Press, New York.

Karpiak, S. E., 1983a, Ganglioside treatment improves recovery
of alternation behavior after unilateral entorhinal cortex
lesion, Exp. Neurol., 81:330.

Karpiak, S. E., 1983b, Accelerated functional development in the
rat neonate following ganglioside administration, in: abstract book "The Cell Biology of Neuronal Plasticity," 6-10
June, Villasimius.

Kasarskis, E. J., Karpiak, S. E., Rapport, M. M., Yu, R. K., and
Bass, N. H., 1981, Abnormal maturation of cerebral cortex and
behavioral deficits in adult rats after neonatal administration of antibody to ganglioside, Dev. Brain Res., 1:25.

Klenk, E., 1942, Uber die Ganglioside, eine neue Gruppe von
zuckerhaltigen   Gehirnlipoiden,   Hoppe Seylers Z. Physiol.
Chem., 273:76.

Ledeen, R. W., 1978, Ganglioside structures and distribution: are
they localized at the nerve ending?, J. Supramol. Struct.,
8:1.

Leon, A., Facci, L., Benvegnù, D., and Toffano, G., 1982, Morphological and biochemical effects of gangliosides in neuroblastoma cells, Develop. Neurosci., 5:108.

Levi-Montalcini, R., 1982, Developmental neurobiology and the
natural history of nerve growth factor, Ann. Rev. Neurosci.,
5:341.

Manthorpe, M., Nieto-Sampedro, M., Skaper, S. D., Lewis, E. R.,
Barbin, G., Longo, F. M., Cotman, C. W., and Varon, S., 1983,
Neuronotrophic activity in brain wounds of the developing
rat. Correlation with implant survival in the wound cavity.
Brain Res., 267:47.

Morgan, J. I., and Seifert, W., 1979, Growth factors and gangliosides: a possibile new perspective in neuronal growth
control, J. Supramol. Struct., 10:111.

Nieto-Sampedro, M., Lewis, E. R., Cotman, C. W., Manthorpe, M.,
Skaper, S. D., Barbin, G., Longo, F. M., and Varon, S., 1983,
Brain injury causes a time-dependent increase in neuron
trophic activity at the lesion site, Science, 217:860.

Oderfeld-Nowak, B., Jezierska, M., Ulas, J., Skup, M., Mitros, K.,
and Wieraszko, A., 1983, Plastic responses of cholinergic
parameters in the hippocampus induced by entohirnal cortex
lesions are intensified by GM$_1$ ganglioside treatment, in: abstract book "The Cell Biology of Neuronal Plasticity", 6-10
June, Villasimius.

Pfenninger, K. H., and Maylie-Pfenninger, M.-F., 1978, Characterization, distribution and apperance of surface carbohydrates
on growing neurites, in: "Neuronal Information Transfer," A.

Kerlin, H. J. Vogen and V. M. Tennyson, eds., Academic Press, New York.

Purpura D. P., and Baker, H. J., 1978, Meganeurites and other aberrant processes of neurons in feline GM_1-gangliosidosis: A Golgi study, Brain Res., 143:13.

Roisen, F. J., Bartfeld, H., Nagele, L., and Yorke, G., 1981, Gangliosides stimulation of axonal sprouting in vitro, Science, 214:577.

Schwartz, M., and Spirman, N., 1982, Sprouting from chicken embryo dorsal root ganglia induced by nerve growth factor is specifically inhibited by affinity-purified antiganglioside antibodies. Proc. Natl. Acad. Sci. USA, 79:6080.

Sharom, F. J., and Grant, C. W. M., 1978, A model for ganglioside behaviour in cell membranes, Biochim. Biophys. Acta, 507:280.

Spirman, N., Sela, B. A., and Schwartz, M., 1981, Antiganglioside antibodies inhibit neuritic outgrowth from regeneration goldfish retinal explants, J. Neurochem, 39:874.

Svennerholm, L., 1963, Chromatographic separation of human brain gangliosides, J. Neurochem, 10:613.

Thoenen, H., and Barde, Y.-A., 1980, The physiology of nerve growth factor, Physiol. Rev., 60:1284.

Thoenen, H., Barde, Y.-A., and Edgar, D., 1982, Factor involved in the regulation of the survival and differentiation of neurons, in: "Repair and Regeneration of the Nervous System", J. G. Nicholls, ed., Springer-Verlag, Berlin.

Toffano, G., Savoini, G., Moroni, F., Lombardi, G., Calzà, L., and Agnati, L. F., 1983, GM_1 ganglioside stimulates the regeneration of dopaminergic neurons in the central nervous system, Brain Res., 261:163.

Varon, S., and Adler, R., 1980, Nerve growth factors and control of nerve growth, in: "Current Topics in Developmental Biology, Vol. 16: Neural Development pt. 2, Neural Development in Model Systems," R. K. Hunt, A. A. Moscona and A. Monroy, eds., Academic Press, New York.

Varon, S., and Adler, R., 1981, Trophic and specifying factors directed to neuronal cells, in: "Advances in Cellular Neurobiology, Vol. 2," S. Fedoroff and L. Hertz, eds., Academic Press, New York.

Varon, S., Manthorpe, M., Selak, I., and Skaper, S. D., 1983, Humoral agents modulating neuronal survival in vitro, in: "Neuromodulation and Brain Function," Pergamon Press, Oxford, in press.

Willard, M., and Skene, J. H. P., 1982, Molecular events in axonal regeneration, in: "Repair and Regeneration of the Nervous System", J. G. Nicholls, ed., Springer-Verlag, Berlin.

Willinger, M., and Schachner, M., 1980, GM_1 ganglioside as a marker for neuronal differentiation in mouse cerebellum, Develop. Biol., 74:101.

Yamakawa, T., and Nagai, Y., 1978, Glycolipids at the cell surface and their biological functions, TIBS, 3:128.

CONCLUDING REMARKS

I must confess that, when I was first shown the program planned by the organizers of this conference, I was left somewhat bewildered to see the apparently unrelated collection of scientific contributions which were to be presented.

However, by the end of these two exciting days, it became clear that the unifying theme of this meeting was that of the plasma membrane and its interaction with peptide hormones, proteins and growth factors and that such an interaction is of fundamental importance in regulating and/or modulating the metabolism and growth of the eukaryotic cell.

Although these conclusions are not new, the examples which have been cited by the numerous specialists and investigators in this field allow us to clearly delineate a common pattern in the apparently complicated array of natural models and experimental systems described to us.

It is not only my contention, but that of others, that in the next twenty years the cell membrane will acquire a place of importance in modern biology similar to that held for the last thirty years by studies on the nucleus which have been focused on understanding the structure and function of the genetic material. This meeting is typical of the beginning of this new age in cellular and molecular biology. Altough several apparently unrelated topics were brought to our attention, they were nevertheless bound by a definite common denominator.

It was appropriate to begin the meeting with one or two papers dealing with the kinetics of the interaction between the cell membrane and molecules such as hormones, growth factors, ligands, etc.; Drs. Minton and Strom have provided us with a clear summary of the theoretical tools needed to study such interactions.

One of the principal mechanisms responsible for the internalization of effectors and ligands is the "receptor-mediated endocytosis". Dr. Edelhoch has reported on the functional and chemical properties of one component of the endocytotic system, namely the coated vesicles and their principal component, the clathrin. Even though alternative models have been proposed by other leading investigators in this field such as Dr. Ira Pastan and coworkers, the coated vesicles seem to be an essential part of the internalization process for hormones and other ligands.

Ionic movements across the cell membrane are clearly regulated by the structure and function of the plasma membrane: quite a few interesting papers on this type have been presented by various groups of investigators among which the French colleagues at Hopital Necker (in collaboration with Dr. Verna), Dr. Moolenaar et al. from the Netherlands, etc.

The adenylate cyclase-cAMP system is also involved in several experimental models: the hormone activation of the enzyme by glucocorticoids (Tolone and coworkers), the mitogenic activity of thyrotropin (Ambesi-Impiombato et al.), the TSH-receptor mediated growth of thyroid cells (Kohn et al.), the growth regulation in S49 mouse lymphoma (Hochman and Levi), in a CHO cell line transformed by the Rous Sarcoma Virus (Gottesman et al.), the response of heart cells to polyamines (Caldarera et al.).

The interaction of membranes with peptides (such as the opioid dermorphins) or with very large and complex proteins (such as thyroglobulin) and small molecules (such as neurotransmitters) have also been the object of very interesting reports.

I could continue to enumerate all of the interesting contributions presented but I don't see the point and my lack of mentioning them certainly does not detract from their importance. The lively and clarifying discussion which followed each presentation is a proof of their high scientific interest.

The success of this meeting should represent and constitute a pledge for the organizers to allow us to enjoy such an enthusiastic experience the coming year. It is with wish that I, on behalf of all the participants, thank the organizers, and, on behalf of the organizers, thank all the participants that have brought their experimental results and scientific wisdom to this successful and friendly gathering.

Gaetano Salvatore M.D.
Professor of General Pathology
Dean, II Faculty of Medicine
University of Naples, ITALY

CONTRIBUTORS

ALOJ S.
Lab. of Biochemical Pharmacology
NIADDKD - NIH.
Bethesda Md. - USA

AMBESI IMPIOMBATO F.S.
Centro Endocrinologia ed
Oncologia Sperimentale C.N.R.
Univ. di Napoli - Italy

BENVEGNU' D.
FIDIA Research Laboratories
Abano Terme - Italy

BIONDI C.
Ist. di Fisiologia Generale
Univ. di Ferrara - Italy

BLASI F.
Ist. di Genetica e Biofisica CNR
Napoli - Italy

BOLIS C.L.
Ist. di Fisiologia Generale
Univ. di Messina - Italy

BONASERA L.
Ist. di Patologia Generale
Univ. di Palermo - Italy

BORASIO P.G.
Ist. di Fisiologia Generale
Univ. di Ferrara - Italy

BORGHETTI A.F.
Ist. di Patologia Generale
Univ. di Parma - Italy

BRAQUET P.
IHB Lab.
Le Plessis Robinson - France

BROCCARDO M.
Ist. di Farmacologia
Univ. di Roma - Italy

CALDARERA C.M.
Ist. di Chimica Biologica
Univ. Bologna - Italy

CAPPUZZO A.
Ist. di Fisiologia Generale
Univ. di Ferrara - Italy

CARAMIA F.G.
Ist. di Patologia Generale
Univ. di Roma - Italy

CLO' C.
Ist. di Chimica Biologica
Univ. di Bologna - Italy

CONSIGLIO E.
Centro Endocrinologia e
Oncologia Sperimentale CNR
Univ. Napoli- Italy

CONTI M.
Ist. di Istologia ed Embriologia
Univ. Roma - Italy

CUGINI P.
Ist. Patologia Generale
Univ. Roma - Italy

DAL TOSO R.
FIDIA Research Lab.
Abano Terme - Italy

DE FEUDIS F.V.
IHB Laboratories
Le Plessis Robinson - France

DE LAAT S.W.
Hubrecht Lab.
Utrecht - The Netherlands

DE LUCA M.
Lab. Biochemical Pharmacology
NIADDKD - NIH
Bethesda Md. - USA

DIEZ J.
INSERM U7 Hopital Necker
Paris - France

EDELHOCH H.
Clinical Endocrinology Branch
NIADDK- NIH
Bethesda Md. - USA

FABBRI E.
Istituto di Fisiologia Generale
Univ. di Ferrara - Italy

FACCI L.
FIDIA Research Lab.
Abano Terme - Italy

FERRETTI E.
Ist. di Fisiologia Generale
Univ. Ferrara - Italy

FORMISANO S.
Centro Endocrinologia e
Oncologia Sperimentale CNR
Univ. Napoli - Italy

FRATI L.
Ist. Patologia Generale
Univ. Roma - Italy

GALEFFI P.
Ist.Genetica e Biofisica CNR
Napoli - Italy

GARAY R.
INSERM U7 Hopital Necker
Paris - France

GEREMIA R.
Ist. di Istologia ed Embriologia
Univ. Roma - Italy

GOTTESMAN M.M.
Lab. of Molecular Biology NIH
Bethesda Md. - USA

GROLLMAN E.
Lab. Biochemical Pharmacology
NIADDKD - NIH
Bethesda Md. - USA

HOCHMAN J.
Dept. of Genetics
Univ. Jerusalem - Israel

IMPROTA G.
Ist. di Farmacologia
Univ. Roma - Italy

JOOSEN L.
Hubrecht Laboratory
Utrecht - The Netherlands

KAJTANIAK-LOCATELLI E.
Ist. Genetica e Biofisica CNR
Napoli - Italy

KOHN L.D.
Lab. of Biochemical Pharmacology
NIADDKD - NIH
Bethesda Md. - USA

LEON A.
FIDIA Research Lab.
Abano Terme - Italy

LEVI E.
Dept. of Zoology
Univ. Jerusalem - Israel

LO PRESTI P.
Ist. Patologia Generale
Univ. Palermo - Italy

LULY P.
Dept. di Biologia
II Univ. ROMA - Italy

MARCOCCI C.
Lab. Biochemical Pharmacology
NIADDKD NIH
Bethesda Md - USA

MARMIROLI S.
Ist. di Chimica Biologica
Univ. Bologna - Italy

MELCHIORRI P.
Ist. di Farmacologia
Univ. Roma - Italy

MINTON A.
Lab. of Biochemical Pharmacology
NIADDKD - NIH
Bethesda Md. - USA

MONACO L.
Ist. di Istologia ed Embriologia
Univ. Roma - Italy

MOOLENAAR W.H.
Hubrecht Lab.
Utrecht - The Netherlands

NANDI P.K.
Clinical Endocrinology Branch
NIADDK - NIH
Bethesda Md - USA

NAZARET C.
INSERM U7 Hopital Necker
Paris - France

NEGRI L.
Ist. di Farmacologia
Univ. Roma - Italy

PETRONINI P.G.
Ist. di Patologia Generale
Univ. di Parma - Italy

PICONE R.
Centro Endocrinologia ed
Oncologia Sperimentale CNR
Univ. Napoli - Italy

PIEDIMONTE G.
Ist. Patologia Generale
Univ. Parma - Italy

PIGNATTI C.
Ist. Chimica Biologica
Univ. Bologna - Italy

ROSSI P.
Ist. Istologia ed Embriologia
Univ. Roma - Italy

ROTH C.
Lab. Molecular Biology - NIH
Bethesda Md. - USA

SALVATORE G.
Ist. di Patologia Generale
Univ. Napoli - Italy

SHIFRIN S
Lab. Biochemical Pharmacology
NIADDKD - NIH
Bethesda Md - USA

SILVOTTI L.
Ist. Patologia Generale
Univ. Parma - Italy

STEFANINI M.
Ist. Istologia Embriologia
Univ. Roma - Italy

STOPPELLI M.P.
Ist. di Genetica e Biofisica   CNR
Napoli - Italy

STROM R.
Dipt. Biopatologia Umana
Univ. Roma - Italy

TOFFANO G.
FIDIA Research Lab.
Abano Terme - Italy

TOLONE G.
Ist. Patologia Generale
Univ. Palermo - Italy

TOSCANO M.V.
Ist. Istologia ed Embriologia
Univ. Roma - Italy

TRAMACERE M.R.
Ist. Patologia Generale
Univ. Parma - Italy

TRAMONTANO D.
Centro Endocrinologia ed
Oncologia Sperimentale CNR
Univ. Napoli - Italy

TREVISANI A.
Ist. Fisiologia Generale
Univ. Ferrara - Italy (deceased)

VALENTE W.
Lab. Biochemical Pharmacology
NIADDKD-NIH
Bethesda Md. - USA

VENEZIANI B.M.
Centro Endocrinologia ed
Oncologia Sperimentale CNR
Univ. Napoli - Italy

VERDE P.
Ist. CNR di Genetica e Biofisica
Napoli - Italy

VERNA R.
Ist. Patologia Generale
Univ. Roma - Italy

VILLONE G.
Centro Endocrinologia ed
Oncologia Sperimentale CNR
Univ. Napoli - Italy

VLAHAKIS G.
Lab. of Molecular Biology - NIH
Bethesda Md. - USA

INDEX